Take the POWER back!

A journey to Electrical Autonomy

Emmanuel Fragnière
Institute of Tourism
School of Management Valais-Wallis (HES-SO)
Switzerland

George Sullivan
Antenna Research Associates - Maryland
USA

Loïc Viret
Evaarist SA - Autarcie.ch, Sion
Switzerland

CRC Press
Taylor & Francis Group
Boca Raton London New York

CRC Press is an imprint of the
Taylor & Francis Group, an **informa** business
A SCIENCE PUBLISHERS BOOK

Cover illustration courtesy of James J Choi

First edition published 2026
by CRC Press
2385 NW Executive Center Drive, Suite 320, Boca Raton FL 33431

and by CRC Press
4 Park Square, Milton Park, Abingdon, Oxon, OX14 4RN

CRC Press is an imprint of Taylor & Francis Group, LLC

Library of Congress Cataloging-in-Publication Data (applied for)

ISBN: 978-1-032-65491-1 (hbk)
ISBN: 978-1-032-68982-1 (pbk)
ISBN: 978-1-032-68984-5 (ebk)

DOI: 10.1201/9781032689845

Typeset in Times New Roman
by Prime Publishing Services

Preface

In an era marked by accelerating climate change, geopolitical instability, and growing energy insecurity, the question of how we power our lives has never been more urgent. Take the POWER Back! is a call to action and a practical guide for individuals, communities, and institutions seeking energy autonomy.

The book emerged from a shared conviction that energy autonomy is possible and essential. Drawing on decades of experience in energy systems, sustainability, and community engagement, we explore the transformative potential of microgrids—localized energy networks that empower users to produce, store, and manage their own electricity. Through a blend of technical insight, economic analysis, and real-world case illustrations, we illuminate the path toward resilient, decentralized, and low-carbon energy systems.

Our goal is to demystify the complexities of energy infrastructure and provide readers with the knowledge necessary to make informed decisions. Whether you are a policymaker, entrepreneur, engineer, student or concerned citizen, this book offers a roadmap to navigate the energy transition with confidence and clarity.

We are grateful to the many communities, researchers, and innovators whose work inspired this project. Their stories, challenges, and successes are woven throughout these pages, offering cautionary lessons and hopeful visions.

Ultimately, Take the POWER Back! is about more than technology. It's about empowerment, equity, and the collective will to shape a sustainable future. We invite you to join us on this journey.

Emmanuel Fragnière
George Sullivan
Loïc Viret

Foreword

By Lisa Jacobson

Imagine a small coastal school, recently battered by a major hurricane. While surrounding neighborhoods went dark for days, this school remained fully operational—its lights powered, food refrigerated, and emergency communications intact. Why? Because it had installed a microgrid: a localized, resilient energy system designed to operate independently when the central grid goes down.

Scenarios like this are no longer exceptional—they are increasingly essential. Around the country and across the globe, communities are facing the stark realities of an aging centralized grid that struggles under the weight of extreme weather, surging demand, and growing digital dependence. In that context, microgrids and distributed energy resources are emerging not just as technical innovations, but as lifelines.

Take the Power Back arrives at exactly the right moment. It captures the urgency of this transformation and offers a practical, inspiring roadmap to a more resilient, equitable, and sustainable energy future.

Our current energy infrastructure, once a symbol of modern progress, is now under strain. The traditional centralized grid—designed for economies of scale and long-distance power delivery—is increasingly vulnerable to cyber threats, natural disasters, and capacity shortfalls. The 2025 *Sustainable Energy in America Factbook,* published by the Business Council for Sustainable Energy and Bloomberg NEF, underscores this trend: electricity demand in the U.S. surged to a 20-year high in 2024, driven by electrified buildings, transportation, increased manufacturing, and the growth of artificial intelligence and data centers. AI, while transformative, comes with an energy footprint—and we must be ready to meet that demand without compromising resilience or sustainability.

This is where microgrids offer a compelling solution. As this book expertly details, microgrids are localized energy systems that can operate autonomously or in coordination with the larger grid. They integrate diverse energy sources—

Lisa is the President of the Business Council for Sustainable Energy, a Washington DC based 70-member business coalition of clean technology companies and trade associations that advocate for policies that support a diverse portfolio of energy options in the United States and ensure access to affordable, reliable, and clean energy solutions. She shares these perspectives in her personal capacity.

including solar, wind, battery storage, and efficient fueled generation—to serve the specific needs of a site, community, or campus. Their ability to island from the grid during emergencies makes them ideal for resilience. And their flexibility allows for cleaner, smarter energy use tailored to real-time demand.

The *2025 Factbook* provides ample evidence of this momentum: 317 gigawatts of clean energy capacity entered the interconnection queue last year, and distributed energy technologies—especially solar, storage, and microgrids—continue to accelerate. Utilities, private sector innovators, and public agencies are investing heavily in these systems, recognizing their potential to stabilize the grid, reduce emissions, and empower end users.

Artificial intelligence is both a driver of this change and a tool to manage it. AI can optimize microgrid performance, forecast demand, and balance supply in real time. It helps manage the complexity of integrating intermittent renewables with storage and demand response, making microgrids more intelligent, efficient, and scalable.

Beyond the technology, *Take the Power Back* makes a powerful case for the human dimension of this energy transition. Microgrids shift the dynamic from passive consumption to active participation. They empower "prosumers"—individuals and communities that produce and manage their own energy—and foster a more democratic, locally driven energy economy. For underserved communities, microgrids can mean access where the traditional grid falls short. For businesses, they offer protection against price spikes and outages. For everyone, they represent a tangible step toward climate action and energy justice.

Importantly, this book does not gloss over the challenges. It addresses head-on the financial, regulatory, and logistical hurdles that stand in the way of widespread adoption. It explores real-world case studies—like Taylor Farms' off-grid facility in California—that demonstrate not only the feasibility but the value of microgrids. It also provides a nuanced look at the policy tools and incentive structures that can help scale these systems, making the book a resource not just for advocates, but for practitioners, policymakers, and planners.

For readers new to the topic, this book offers clarity. For those already working in the field, it offers momentum. And for all of us—whether we are concerned about rising electricity bills, grid reliability, or climate change—it offers a hopeful, actionable vision for the future.

As someone who has spent my career working to accelerate clean, affordable, and reliable energy, I find *Take the Power Back* to be both timely and empowering. It reminds us that the energy transition is not just a technical shift—it's a societal one. And it puts the tools for change directly in the hands of those ready to act.

Whether you're a policymaker, a business leader, a community advocate, or a curious homeowner, I hope this book sparks ideas and inspires confidence. The future of energy will be decentralized, resilient, and increasingly shaped by people like you.

Now is the time to take the power back!

Executive Summary

In a world facing unprecedented environmental, political, and social challenges, energy autonomy is quickly emerging as a vital goal for individuals, communities, and nations alike. The traditional model of centralized energy production and distribution is increasingly seen as inadequate to address the pressing needs of the 21st century. This book delves into the opportunities to achieve energy self-sufficiency via microgrids and offers a guide to understanding and implementing decentralized energy systems.

The Problem with Centralized Energy Systems

Centralized energy systems, which have dominated global energy production, distribution and end use since the industrial revolution, are built on the premise of efficiency, affordability and reliability being driven by economies of scale—large power plants generating electricity that is transmitted over vast distances to a large number and variety of end users early on proved to be the most effective and efficient technoeconomic model. While effective in their time, these systems are now showing signs of strain. Aging infrastructure, inefficiencies in transmission, and a growing vulnerability to natural disasters and cyber-attacks have exposed the limitations of relying on an interconnected grid or grids. Furthermore, the increasingly apparent environmental and climatic impacts of large-scale fossil fuel combustion, which remains the primary generation method for most of these systems, is providing a strong sense of urgency to the efforts to transformation to renewable and carbon-free sources of power generation.

The Rise of Microgrids and Storage: A Key Aspect of Future Power Security and Sustainability

Microgrids offer a promising solution to many of these challenges. A microgrid is a localized energy system capable of operating independently or in concert with the larger grid. By offering the possibility of integrating a variety of energy sources, including solar, wind, hydroelectric and fueled power generation at a small and distributed scale closer to or co-located with the end users, microgrids provide a more sustainable, resilient, and reliable alternative to traditional energy systems. The decentralized nature of microgrids reduces reliance on distant power plants and long transmission lines, minimizing the risk of widespread outages and enhancing energy security.

Key Themes and Structure of the Book

The book is structured to guide the reader through the multifaceted world of microgrids, beginning with the motivations for pursuing energy autonomy and culminating in practical strategies for implementing microgrids in various contexts.

1. Understanding the Shift to Energy Autonomy

The initial chapter explores the driving forces behind the transformation to microgrids. The authors examine economic factors, such as the rising cost of energy and the desire for price stability, alongside environmental concerns, particularly the need to reduce greenhouse gas emissions. The authors also explore societal motivations, such as the aim for local energy independence and efforts to ensure fair access to energy in disadvantaged communities.

2. Technical Foundations of Microgrids

A deep dive into the technical aspects of microgrids follows, providing a detailed overview of the components and technologies that make microgrids viable. This includes analyzing the range of generation sources, energy storage solutions, and smart grid technologies. The book explains how these elements work together to create a self-sustaining energy system that can adapt to varying demands and conditions.

3. Economic Considerations and Business Models

In this chapter, the authors explore the financial challenges of implementing microgrids. While the initial investment in microgrid infrastructure can be significant, the long-term benefits—such as reduced energy costs, improved energy security, and the potential for generating revenue through the valorization of excess power—make microgrids an attractive option in many places. Various business models, including public-private partnerships and community ownership, are discussed as means to make microgrids accessible and financially viable.

4. Social and Environmental Impacts

This chapter looks at the future of microgrids. It discusses new trends, potential problems, and their impact on society and the environment. New technology will make microgrids more efficient and help them spread faster. However, the transition to smaller energy systems will face problems like regulations and upfront costs. Beyond the technical and economic aspects, the book also delves into the broader social and environmental implications of microgrids. Microgrids have the potential to democratize energy, giving individuals and communities greater control over their energy supply and reducing dependence on large or monopolistic energy providers whose reliability is increasingly challenged. This shift can lead to better and wider access to reliable energy, particularly in remote areas and communities currently experiencing below average power availability. Environmentally, microgrids help reduce harmful emissions by utilizing renewable or efficient fueled power generation, thereby helping to mitigate the impacts of climate change and improving the quality of the environment.

5. Hypothetical Scenarios and Practical Applications

The book presents several examples that illustrate the diverse applications of microgrids around the world. These highlight the flexibility of microgrids, showing how they can be tailored to meet the specific needs of different communities, whether in urban or rural settings, developed or developing countries. Each example provides insights into the challenges faced and the solutions implemented, offering valuable lessons for readers considering microgrid projects.

6. Conclusion: The Path Forward

By adopting microgrids and pursuing energy self-sufficiency, we can build a more sustainable, resilient, and available energy system—one that meets the needs of today without compromising the ability of future generations to meet their own.

TAKE THE POWER BACK is a guide to energy autonomy for anyone interested in understanding and contributing to the energy transformation. Whether you are a policy maker, business leader, engineer, or regular citizen looking to have more reliable and, if desired, zero emissions power, this book provides the knowledge and tools needed to navigate the complexities of energy autonomy and to play a part in shaping the future of energy.

Contents

Preface	*iii*
Foreword	*v*
Executive Summary	*vii*
Introduction	*xiii*
1. The Why: Is Energy Autonomy Right for You?	1
2. The What: Building Blocks of Energy Independence	30
3. How Do We Implement a Microgrid?	85
4. Who Leads the Energy Evolution and What's Next?	118
5. Greenberg Family's Journey through Energy Autonomy	165
Glossary	191
Index	197
Biography	199

Introduction

In an era defined by rapid technological advancement and growing environmental consciousness, the quest for energy autonomy has emerged as a critical imperative. The traditional centralized energy systems, once lauded for their ability to power nations and drive industrial growth, are now increasingly seen as relics of a bygone era—unsustainable, vulnerable, and ill-suited to the needs of a changing world. The rise of microgrids represents a paradigm shift in how we generate, distribute, and consume energy, offering a pathway to self-sufficiency, resilience, and sustainability.

This book is an exploration of that shift— a journey into the world of microgrids and the broader concept of energy autonomy. The term "energy autonomy", often associated with economic self-sufficiency, is here applied to the realm of energy, where it signifies the ability of individuals, communities, and businesses to produce and manage their own electricity independently of large-scale, centralized grids. This movement towards energy independence is not just a technical challenge but a response to a host of interconnected global issues: the escalating costs of energy, environmental degradation caused by fossil fuel exploitation and consumption, the vulnerability of centralized grids to natural disasters and cyber attacks, and the growing demand for reliable and zero emission energy sources.

To understand the significance of this shift, we must first delve into the history of electricity and the evolution of our current energy systems. The centralization of energy production, which began with the advent of large-scale power plants in the early 20th century, was driven by economies of scale and the need to deliver electricity to an ever-growing population. However, this model, while successful for decades, has inherent weaknesses. The massive infrastructure required to generate and transmit electricity across vast distances is costly to maintain, prone to inefficiencies, and vulnerable to disruptions. Moreover, widespread fossil fuel consumption is contributing to climate change and threatening some of the planet's ecosystems.

Microgrids offer a compelling alternative. These localized energy systems can operate independently or in conjunction with the larger grid, providing a reliable and sustainable source of power tailored to the specific needs of the community or business they serve. By being able to integrate a wide range of generation types, including renewable energy sources such as solar, wind, and hydroelectric power, microgrids can reduce dependence on fossil fuel supply and help mitigate the effects of climate change. Furthermore, they enhance energy security by decentralizing power production and reducing the risk of large-scale outages.

Microgrids offer direct economic benefits as well. They generate power locally, protecting against price volatility and can be aggregated into virtual power plants. They also create jobs in modern energy technologies and can provide new revenue and tax streams.

This book is structured to provide the reader with a comprehensive understanding of microgrids and their potential to revolutionize the way we think about energy. We begin by examining the motivations behind the shift towards energy autonomy, exploring the economic, environmental, and social drivers that are pushing individuals and communities to seek greater control over their energy supply. We then delve into the technical aspects of microgrids, discussing the various components, technologies, and systems that make them viable. This includes an exploration of the variety of energy sources available now and expected in the near future, energy storage solutions, and smart grid technologies that enable efficient management of energy production and consumption.

Examples from around the world are presented to illustrate the diverse applications and benefits of microgrids. From rural communities in developing countries seeking to electrify remote areas, to urban neighborhoods and businesses in developed nations looking to enhance resilience against grid failures, these examples demonstrate the flexibility and scalability of microgrid solutions.

The book also addresses the economic considerations of microgrid implementation. While the initial investment can be significant, the long-term savings in energy costs, coupled with the environmental and social benefits, often justify the expense. In addition to the technical and economic aspects, we consider the social implications of the shift towards energy autonomy. Microgrids empower individuals and communities by giving them greater control over their energy supply. This can lead to a more equitable distribution of energy resources, particularly in underserved areas where traditional grid infrastructure is lacking or unreliable. Furthermore, the decentralization of energy production can foster innovation and entrepreneurship, creating new opportunities and jobs for local businesses to supply, maintain and operate microgrids and the technologies needed to support them.

Integrating microgrids can make energy systems more resilient, especially in the face of disasters (natural and human-made). Microgrids are designed to continue working when the main grid is down. This is important for hospitals, emergency services, and communication networks. As we look at more of the benefits of microgrids, we can see that they are not just a new technology. They are a key part of making our energy future more sustainable, fair, and reliable.

Microgrids can also help reduce greenhouse gas and harmful emissions by using renewable energy sources like solar, wind, micro hydroelectric, and biomass fuels. This helps mitigate the effects of climate change and make the environment cleaner.

Finally, microgrids can help strengthen the currently aging electricity distribution infrastructure.

As we navigate through these topics, the overarching theme of this book remains thus: the future of energy can be largely decentralized, autonomous, sustainable, and end user-driven. The increasing deployment of microgrids is not just a technical innovation but also offers the opportunity to move towards a generally more resilient energy system. By the end of this book, you will understand the potential of microgrids to transform our energy landscape and the practical steps you can take to participate in this exciting journey towards energy independence.

1 The Why: Is Energy Autonomy Right for You?

"What's past is prologue." Shakespeare

This chapter makes the case for the urgent need for energy autonomy in a world increasingly affected by environmental crises, rising utility costs, and unreliable centralized power grids. It delves into the economic, social, environmental, and technological motivations that propel individuals, communities, and institutions to pursue electrical self-sufficiency. Readers are guided through the historical evolution of electricity, from its mythical reverence in ancient civilizations to its centralization during the Industrial Revolution. The chapter critically examines the vulnerabilities of the modern grid, including aging infrastructure, inefficiencies in long-distance transmission, and mounting threats from climate-induced disasters and cyberattacks. Through the lens of "cord-cutting" and the rise of prosumers—people who produce and consume electricity—the chapter argues that microgrids offer empowerment and resilience, as well as technological innovation. Real-world examples, such as Taylor Farms in California and Babcock Ranch in Florida, demonstrate how microgrids can enable communities to withstand outages, reduce emissions, and control energy costs. The chapter also addresses the growing dependence on stable electricity in the digital age, driven by IoT devices and artificial intelligence, which further amplifies the need for localized, smart energy systems. Ultimately, this chapter establishes a philosophical and practical foundation by asking readers not only how to achieve energy autonomy but also why they must. The chapter challenges readers to think critically about energy, equity, and the future of sustainability.

Why not stop depending entirely on the electrical grid and move to an "islandable" microgrid? There are many reasons, but one might cite goals of electrical self-sufficiency and economic independence (which can also be regional or even national, depending on the size of the country, if one wishes to contribute to the nation's independence), energy efficiency, long-term cost savings, reduced exposure to utility cost volatility, the need or desire to increase one's energy resilience, improved air quality, and/or reducing greenhouse gas emissions. Some will see other goals such as the social aspect of being able to solve their own problems as a community and to personally create and manage their own energy resources instead of relying on a utility or other outside company or entity to do so (i.e., the social dimension of energy autonomy). However, switching to an islandable microgrid is not easy. It can cost a lot to set up and maintain. It is also difficult to manage and balance energy loads in a system that might use many

DOI: 10.1201/9781032689845-1

different power sources. However, the long-term benefits of energy independence, resilience, and sustainability often outweigh the initial costs for many individuals, communities, and businesses. If you plan carefully and leverage the right incentives and technologies, the switch to a microgrid will make sense for you.

The economic dimension will also play a major role in deciding whether to establish energy autonomy. Utility-supplied power from your grid will often be less expensive per kilowatt hour than power you would obtain from implementing a private microgrid... at least in the short to medium term. The substantial up-front costs of establishing a microgrid can be daunting, or outright prohibitive, to many. In this chapter, we will explain the trends and developments of microgrids in the United States, Europe, and the rest of the world.

Historical Perspective

Before we dive into the relatively new realm of microgrids, the authors thought it might be beneficial to give readers (mainly those new to the subject) a primer on the history of electricity and how the world has arrived at its current state. Electricity has a long history that traces back to ancient times... Let's set the clocks back about, oh, 3,000 years...

In ancient Greece, when the Olympic gods were considered the supreme beings that ruled the world, Zeus (Figure 1.1), the mighty king of Mount Olympus and considered to be the most powerful deity of the Olympic pantheon, wielded bolts of lightning as his weapon of choice. These fiery bolts were not mere weapons; however, they were divine manifestations. But beside Zeus stood Athena, the goddess of wisdom. She wondered: Could there be more to lightning than anger? While human interest in electricity by way of observing the awesome might of lightning surely predated the Greek mythos, this is the first known instance of humanity equating the power to control electricity as being superior to all others. And so began the story of electricity—a force that bridges heaven and earth, from ancient awe to modern power grids.

For the next 2,700 years or so, the mystery of lightning remained the mystical province of the gods. But in 1752, the first conclusive evidence was obtained that lightning is a manifestation of electricity when Thomas Francois-Dalibard, following up on Benjamin Franklin's hypothesis, successfully obtained electricity for the first time using a metal rod during a thunderstorm. Thus, the mystical power of the gods was seized from the heavens and made the plaything of mere human scientists.

It did not take long for the scientific community to learn the uses of and ways to create and control this newfound power (pun intentional). Alessandro Volta invented the voltaic pile—the first battery to generate a continuous electrical current—using a stack of alternating copper and zinc discs separated by brine-soaked cloth that transmitted current continuously through a connected wire. For nearly 70 years this remained the most reliable way to obtain a constant supply of electrical power. While the idea of a dynamo using electromagnetic induction

Fig. 1.1 Zeus (Jupiter) and Lightning Bolt (source: sculpture by James J. Choi) available at (https://jamesjchoi.com/sculptures/196-zeus-jupiter-and-lightning-bolt).

to generate electrical power was first conceived and tested by Michael Faraday in 1831, it took nearly 35 years to develop it to the point where it could be applied at a large scale for power generation. It became the basis for the first generators, and later alternators, used for utility-scale power generation.

From the start, humans' ability to harness electricity was intertwined with advances in how humans communicate. Volta's invention was crucial because it allowed for a consistent and controllable flow of electricity. Additionally, Hans Christian Ørsted's demonstration in 1820 of the connection between electricity and magnetism [1] laid the foundation for Samuel Morse's work on developing the telegraph. Ørsted showed that an electric current could create a magnetic field, which was a pivotal discovery for the development of electromagnets. Building on these principles, Joseph Henry's work on electromagnetism was also instrumental. Henry demonstrated that electrical signals could be transmitted over long distances using a relay system [2]. This system amplified the signal at intervals, overcoming the problem of signal degradation over long wires. Morse, with the help of his colleague Leonard Gale, incorporated these ideas into his telegraph system.

Utilizing these scientific advancements, Morse developed a system where messages moved along wires in a digital code of dots and dashes. This "lightning messenger" crossed continents in seconds, enabling lovers to share secrets, stock markets to react almost instantly, and societal and technological progress to advance rapidly. Today, the telegraph's legacy lives on in the fiber optic cables that crisscross oceans and continents and allow nearly instantaneous communication through texts, e-mails, personal messages, etc.

The Birth of Practical Electricity (Late 1800s)

Both Paris and Timișoara were among the first cities in Europe to use electric street lighting in 1878, based on a direct current (DC) system [3]. However, the invention of the light bulb by Thomas Edison in 1879 marks the most visible beginning of modern electricity. The first fossil fuel power plant in the United States, Pearl Street Station, was built in New York City's financial district to power these light bulbs. Its activation in 1882 marked the advent of the modern electrical grid, enabling the transmission of electricity over long distances. In 1896, the Niagara Falls Power Company demonstrated the use of alternating current (AC) to send electricity over long distances, proving that electricity could power entire cities and regions. Despite these advancements, it took time for households to adopt electric lights. According to the National Park Service, by 1925, only half of the homes in the United States had access to electricity, with the rest relying on kerosene, gas, and candles [4].

There was a spirited debate about whether to change from DC to AC systems. In the late 1880s, Thomas Edison advocated for DC, while Nikola Tesla and George Westinghouse supported AC. AC was easier to convert for different uses and could be transmitted over long distances with fewer losses, which led to its widespread adoption. With the rise of electric lighting, other household appliances such as irons, vacuum cleaners, and washing machines were introduced in the early 20th century, transforming daily life. Electricity also revolutionized industry, making factories more productive and fostering the emergence of new industries.

Electricity changed society. It made the workday longer, safer, and better. In the 1930s, rural electrification programs brought electricity to remote areas, helping the economy. Electrification helped America grow and urbanize. It also helped create the modern world we live in today.

Electricity led to new ideas in many areas. Electric streetcars made cities more accessible and reduced reliance on horse-drawn carriages. Electric elevators let people live in skyscrapers, changing city views and making cities more crowded. Electric telegraphs and telephones let people talk to each other from far away, which improves the economy and people's lives. These new technologies, powered by electricity, helped create the modern world.

American cities set up centralized electrical networks to meet the growing demand for light and power in urban areas. These networks transformed daily life and drove economic growth. However, despite the rapid advances and initial successes, the early electrical networks faced several challenges. Problems with transmission, reliability, and keeping up with urban expansion made it difficult to distribute electricity efficiently and widely. These challenges led to further innovation and investment in infrastructure, which laid the foundation for modern centralized electrical grids.

Electricity was then often viewed as a quasi-magical force. In French, it is referred to as 'la fée électricité' (the electricity fairy) from the painting by Raoul Dufy (Figure 1.2).

In the mid-twentieth century, more and more people wanted electricity. Local grids were struggling to provide adequate supply and adequate quality of supply due to increasing demand and the increasing variety of demands. Centralized grids were thus established to connect different regions of the US and make the power system more efficient. These grids allowed power plants, substations and distribution networks to exchange electricity. This ensured there was enough

Fig. 1.2 La fée électricité (the electricity fairy) from the painting by Raoul Dufy [5].

electricity for everyone; the interconnection promised a more stable power supply that would foster economic growth and technological advancement.

As they developed, interconnected grids became more economically viable. In the late 19th century, people built local grids to serve cities and towns. These early grids were isolated and limited in efficiency and capacity. As technology improved and more people needed electricity, it became clear that a more integrated approach was needed. In the mid-20th century, people started connecting local grids. Regional and national grids let people share electricity across larger areas. Power plants could operate more efficiently because they could supply electricity to multiple regions. This also saved money because each area did not have to build its own power plants.

The development of interconnected grids in the early 20th century made electric power much more reliable and stable. These grids reduce the risk of localized outages by sharing resources and diversifying energy sources. In the event of a storm or other disruption, surplus electricity from neighboring regions can be rerouted to maintain a continuous power supply.

Additionally, the interconnected nature of national grids allows for load balancing and a stable supply of electricity. During peak demand periods, surplus electricity from less-frequented regions can be dispatched to meet increased demand in urban and industrial centers. For example, electricity from hydroelectric plants in a remote area with little to no power demand can be transmitted to a faraway urban area with high demand to help meet demand surges, on a hot summer day when there is a lot of air conditioner use, for example. This helps prevent blackouts and maintain a consistent supply of electricity.

Also, interconnected grids help mix different energy sources, making the power supply more stable and efficient. Some regions use different types of power plants, like coal, hydroelectric, or natural gas-fired plants. Connecting these sources makes the grid more reliable and makes it less likely that there will be widespread power outages even if one power station is temporarily unavailable. By allowing regions to share resources and support each other during disruptions, these grids made it possible for us to have the electrical systems we have today.

Interconnected grids are also good for the economy. They use less energy and are cheaper to run. Upgrades to these systems have made them more efficient and cost-effective. They can also adapt to meet the growing energy needs of expanding urban areas. This was the established paradigm from the 1970s until the present. In the 1970s, grids became even more interconnected. During this time, money was spent to improve the way electricity is sent and managed. Upgrades included high-voltage transmission lines, which let electricity travel longer distances with less loss. This helped distribute electricity from areas with more power to places that need it.

As cities grew and industry boomed, the grids proved adaptable. They could supply cities with more electricity from areas with extra capacity. This was important during periods of rapid urbanization, as it ensured new developments had a reliable power supply. The ability to balance loads across a wide area also

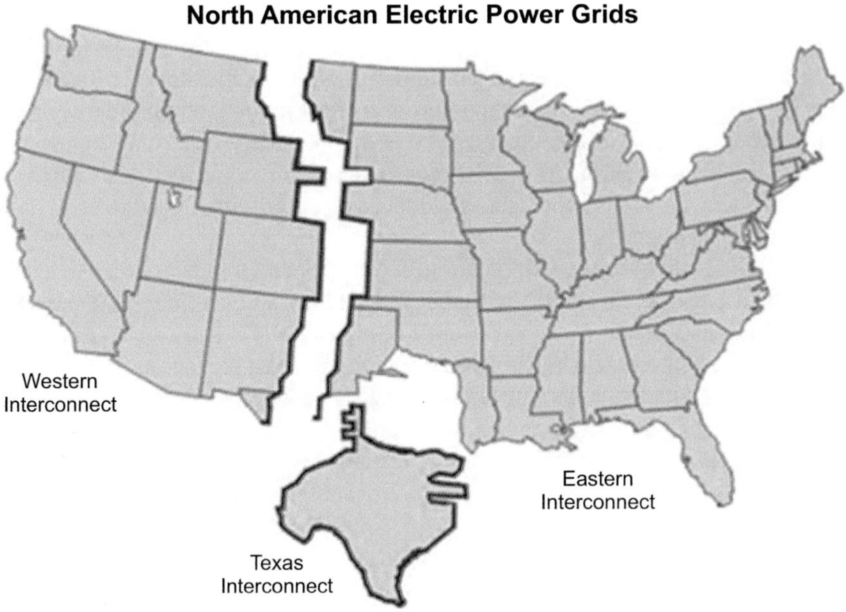

Fig. 1.3 North American Electric Power Grids and Market. Source: EPA, the main energy grids in the US [6].

helped prevent blackouts. The idea of interconnected grids continued to change throughout the 20th and 21st centuries. New technology made the grids even more efficient and reliable. Mixing different energy sources like hydroelectric, coal, and natural gas made the energy supply more balanced and stable. This made the power system more stable and less likely to change prices.

The North American electric power grids have evolved to be reliable, efficient, and cost-effective. The US has three major regional grids (Figure 1.3) the Eastern Interconnection, the Western Interconnection, and the Texas Interconnection (ERCOT) [7]. In the early 1900s, US utilities started working together to share power. This led to networks that were connected. By the 1960s, most of the continental U.S. was covered by two major grids: the Eastern and Western Interconnections. The Texas Interconnection operates mostly independently due to its unique regulatory environment. The Energy Policy Act of 1992 further shaped the grid by promoting competition and requiring transmission line owners to provide open access to their networks. This restructuring allowed for more efficient and reliable electricity distribution across the country. Today, these three interconnections function mostly independently, with limited power exchanges between them.

Europe's national electricity grids have developed differently because of the continent's diverse political, economic, and geographical landscape. Europe has many national grids that are connected through a network of cross-border transmission lines. The European grid system started as individual national grids. These grids were connected to make energy more secure, efficient, and renewable. In 2009, the European Network of Transmission System Operators for Electricity (ENTSO-E) was created to make the European electricity market more integrated and coordinated. This integration makes it easier to manage electricity supply and demand across the continent, which helps create a more resilient and sustainable energy system. For information about European electricity grids, see the ENTSO-E Transmission System Map [8].

Following this overview of global grid development trends, we turn to a data analysis based on recent reports by Ember and the International Energy Agency [9]. Global electricity demand reached an all-time high of 29,471 terawatt-hours (TWh) in 2023, a 2.2% increase compared to 2022. Over the past two decades, electricity use has nearly doubled, reflecting major changes in how societies consume energy.

Much of this growth is driven by the ongoing shift toward electrification in heating, cooling, and transport systems. In fact, by 2021, these sectors already accounted for 21% of global final energy consumption. Regional patterns vary widely. Asia now uses more than half (52%) of the world's electricity, though it holds about 55% of the global population—suggesting development potential in other regions. China remains the largest consumer, using 9,441 TWh in 2023, which is 32% of global demand. Since 2000, China's per capita electricity use has risen sixfold, reaching 6.6 MWh per person.

Another major factor in rising electricity demand is the expansion of digital infrastructure, especially artificial intelligence (AI) and data centers. The International Energy Agency projects that data center electricity consumption will more than double by 2030 to around 945 TWh. In 2024, AI servers already accounted for 24% of total server electricity use.

These trends make it clear: to meet this rising demand while also meeting climate goals, countries will need to keep expanding clean electricity generation and upgrading power grids.

Concurrently with this increase in electricity consumption, there has been rapid growth in the adoption and deployment of renewable energy sources, reflecting a global move towards sustainability. The emerging challenge is how to balance rapidly growing electricity demand with sustainable power generation, namely how to harness the potential of renewable energy resources while adopting innovative distribution, storage, and consumption technologies to create a more efficient and environmentally friendly energy landscape.

As the US and other regions of the world begin to face this challenge, individuals, communities, data centers, military installations, hospitals, other businesses, and governments have begun to increasingly desire and, in many cases, plan and take action to ensure their energy resilience and independence. One of the key innovations to enable these efforts was the development of the microgrid concept and related technologies. The transformation of microgrids from expensive, novel systems to commercialized, easily deployable solutions has been facilitated by technological advances such as highly efficient, cost-effective solar panels, widely available battery energy storage and sophisticated control systems that allow for decentralized energy production and utilization. This transformation was not only an organic technological and industrial evolution but also prompted in large part by national security, industrial safety, economic and social pressures, and considerations.

In the early 21st century, the application of microgrid technology began to spread in the United States. Consumers in California, for example, were simultaneously incentivized by both environmental sustainability considerations and increasing grid instability caused by extreme fire and weather events. Communities like Borrego Springs, Blue Lake Rancheria, and the Sonoma County Public Safety Power Shutoff Project show how effective microgrids are. They provide power during outages using solar and genset generation and battery storage. In Florida, the increasing deployment of microgrids by the largest utility, FPL, is a direct response to more frequent outages caused by increasingly frequent, and more powerful, hurricanes and tropical storms. For example, Babcock Ranch in Florida used its solar-powered microgrid to stay online constantly during Hurricane Ian, while large swaths of the southern two-thirds of the state went dark for days [10]. Increasingly, the adoption of microgrids extends beyond technical considerations and represents a broader movement toward economic and social empowerment. Energy communities are investing in local energy solutions, motivated by a desire for energy self-sufficiency and economic independence.

To illustrate this, let's present here such a case of social empowerment and the notion of "cord cutting", i.e., when people cancel their cable or satellite TV subscriptions and use only internet-based services instead. People are cutting the cable because they want more flexible and cheaper ways to access content. In the context of energy, as illustrated by the Taylor Farms example below [11], cord cutting can also entail disconnecting from the traditional power grid to achieve energy independence through alternative sources like solar power and fuel cells. This approach can provide more reliable and cost-effective energy, especially for large operations where power outages can be extremely costly. In jurisdictions that allow full cord-cutting, you may not have to worry about the interconnection agreement issues and all the regulations and rules that come with that. But in some jurisdictions, you might have to petition for exemption from some "anti cord cutting" laws [12] and ensure you plan accordingly to obtain all applicable permits, inspections, authorizations, etc. to install your system.

As an example of one such cord-cutter, you may look to Taylor Farms, who after suffering unreliable power supply and quality and surging rates from PG&E (Pacific Gas and Electric Company) decided to completely cut the cord and establish energy autarchy [13]. Taylor Farms is a major North American food producer with a 450,000 sq ft packaged salads manufacturing plant in San Juan Bautista, California, USA. In September 2022, Taylor Farms announced it was going off-grid and establishing an eight MW microgrid. Two MW of solar power would address peak load, and 6 MW of natural gas-fueled, solid-oxide fuel cells would provide steady base load supply. In addition, a 2 MW battery would provide additional redundancy in the event of unusually high spikes in load and if some generation assets go offline. It is rumored the CEO of Taylor Farms was so excited about the prospect that he wanted to personally clip the cables to the grid lines. One can hardly blame him when even a relatively brief outage may result in tens or hundreds of thousands of dollars of food being spoiled or deemed unfit for sale.

The Growing Demand for Electricity in the Digital Age

The use of electrical devices like those in the Internet of Things (IoT) is becoming ubiquitous. As our world gets more connected, we need more electricity to establish and maintain those connections. This is especially true in the world where the physical and digital are mixed (i.e., "phygital"). In the past, our daily experiences were mostly physical. Now, they're mostly digital. As we use more smart technology, a reliable and resilient electricity supply will become even more important to maintaining adequate living standards and economic prosperity than it already is. The rise in the popularity of IoT products, reliance on redundant, interconnected communication networks to accomplish some of the most rote tasks, like paying for goods and services or moving money, and the emergence of AI allowing people to outsource much mundane work will drive this trend.

Nearly all modern appliances use chips that use additional electricity, even if it is at a low voltage. For example, not only a smartphone needs electricity to use mobile payment services. It doesn't matter if your phone is fully charged if your Wi-Fi or internet provider is down or your bank's server is offline due to a power outage in one or more of their systems; you will not be able to make a mobile payment. This interface between hardware and the ether is an example of the phygital world. This increasing dependence on the phygital world makes us highly vulnerable to any disruption to power supply, even disruptions that might occur thousands of miles away.

Solving these power supply reliance concerns is not just about replacing fuel with electricity, building more transmission wires, installing more battery backup systems, etc. It is also critical to optimize how we use information, power supply and distribution technology together. The term "smart" is often used for electrical appliances and systems that rely on electronics and computers. But it also applies to AI, which helps these appliances make decisions or helps us make decisions on how to use them most efficiently. Smart technology in appliances and systems is more than just automation and connectivity. Many "smart" devices currently rely on simple algorithms and are not AI-enabled. AI has brought a new era of intelligent automation and decision-making, which when incorporated into IoT devices, does predefined tasks, learns from data, adapts to user behavior, and makes decisions on its own or with user help. Smart appliances will be able to adapt to and anticipate your preferences and thereby use resources more efficiently, making your life easier and optimizing power consumption simultaneously. While the increasing use of AI is expected to drive an overall increase in power demand at the macro level (e.g., proliferation of more power-hungry microprocessors and related data centers to process all the IoT data and AI data processing), there is much opportunity for it to reduce demand at a micro level (e.g., small businesses and residences) by optimizing the functioning of multiple systems and appliances in a home or business.

IoT technology makes these appliances more connected and useful. IoT devices can talk to each other, collect data in real-time, and respond to changes. Sensors, actuators, and network connectivity create this interconnected ecosystem. This makes it easy to connect appliances and systems in a smart home. For example, a smart thermostat with AI learns your habits. It adjusts the temperature to save energy and keep you comfortable. Sensors in your home detect when you're there, the light, and the weather, whether other appliances in the home are using power or generating heat. This helps the thermostat make better decisions based on what's going on around it and when to operate the cooling/heating systems to avoid demand surges and match the load more closely with renewable power supply and power storage. AI, IoT, and electrical appliances are changing how we interact with technology. Smart, adaptable, and connected devices can make our lives better and help us use resources more wisely.

As we step back to consider the larger historical context, this analysis draws on key data from Ember, The Earth and I, and Visualizing Energy [14]. Global electricity consumption has expanded significantly over the past four decades. In 1980, demand stood at approximately 7,000 TWh. By 2023, it had reached a record high of 29,471 TWh—more than a fourfold increase. This sustained growth reflects a global shift in how energy is produced and consumed.

Acceleration was particularly strong from the late 1990s onward, coinciding with rapid urbanization and industrialization in emerging economies—especially China and India—which became major drivers of electricity demand growth. Looking further back, global electricity generation increased from just 66.4 TWh in 1900 to over 29,000 TWh in 2022, illustrating how electricity evolved from a luxury to a fundamental building block of modern society.

These trends underscore the growing importance of electricity in supporting economic development, digitalization, and improved living standards. Meeting this rising demand will require sustainable policies, investment in clean electricity technologies, and grid modernization to ensure future energy security.

Electricity supply systems have always been developed to meet growing demand and improve reliability. From local grids to national grids, each step has been about ensuring a stable and efficient power supply. Smart grids and IoT technology mark the latest phase in this evolution. These changes are not just about meeting current demand. They are also about anticipating future needs and optimizing the entire energy system. AI and IoT will be key to our future electricity systems. These technologies will help us control and manage energy resources better, reducing waste and improving efficiency. They will help us incorporate renewable energy sources, which fluctuate significantly in output, into the grid. This change primarily refers to the variability in generation. Solar and wind power, for example, depend on weather conditions and are not constant like traditional energy sources. AI has the potential to help us use renewable energy more effectively, which can reduce our reliance on fossil fuels and help mitigate the impacts of climate change. In conclusion, the proliferation of IoT devices and application of AI can make our electricity supply systems more efficient, reliable, and sustainable over time by optimizing our energy use.

The following section will examine the transition from a traditional grid to microgrids, which represents a significant step forward in the evolution of energy distribution. This transition represents a pivotal moment in the history of energy management, as it introduces a more decentralized, resilient, and sustainable approach to the production, distribution, and use of electricity. By examining the evolution and deployment of microgrids, we will elucidate how these distributed energy systems can bolster grid reliability, facilitate the integration of renewable energy sources, and empower communities to assume control of their energy future. This transition addresses the limitations of traditional grids while simultaneously paving the way for a more adaptable and innovative energy landscape. This landscape will be capable of meeting the diverse and growing demands of the 21st century.

From the Traditional Electrical Grid to the Microgrid

The Traditional Electrical Grid

A traditional electrical grid is a centralized network that transports electricity from large production plants to consumers, through high-voltage lines and transformers. It is based on a demand logic, which requires adjusting production according to consumption. Generally, these networks are heavy infrastructures whose beginnings date back to the early 20th century, mainly in developed areas of the world like the US or Central Europe. The means of production are variable, such as nuclear power plants, hydroelectric plants, and gas or coal power plants, as well as new renewable energies as seen before.

Figure 1.4 shows the electricity grid as a complex network that delivers electricity from producers to consumers. Power plants generate electricity from different energy sources like coal, natural gas, nuclear, hydro, wind, and solar. Wind and solar are becoming more widely adopted. Once generated, electricity is sent over long distances to substations. Substations lower the voltage to make it suitable for distribution. From there, electricity travels to homes, businesses, and other users. Transformers make the voltage safe for homes and businesses. Consumers use electricity for things like lighting, heating, cooling, transport and appliances. The electricity grid is important because it makes sure that electricity from power plants reaches consumers safely and efficiently.

A traditional electrical grid is generally vulnerable to failures, energy losses, and voltage fluctuations, and lacks flexibility in terms of its expansion and the integration of decentralized energy production systems. One of the main limiting aspects of a traditional electrical grid is its transportation and distribution

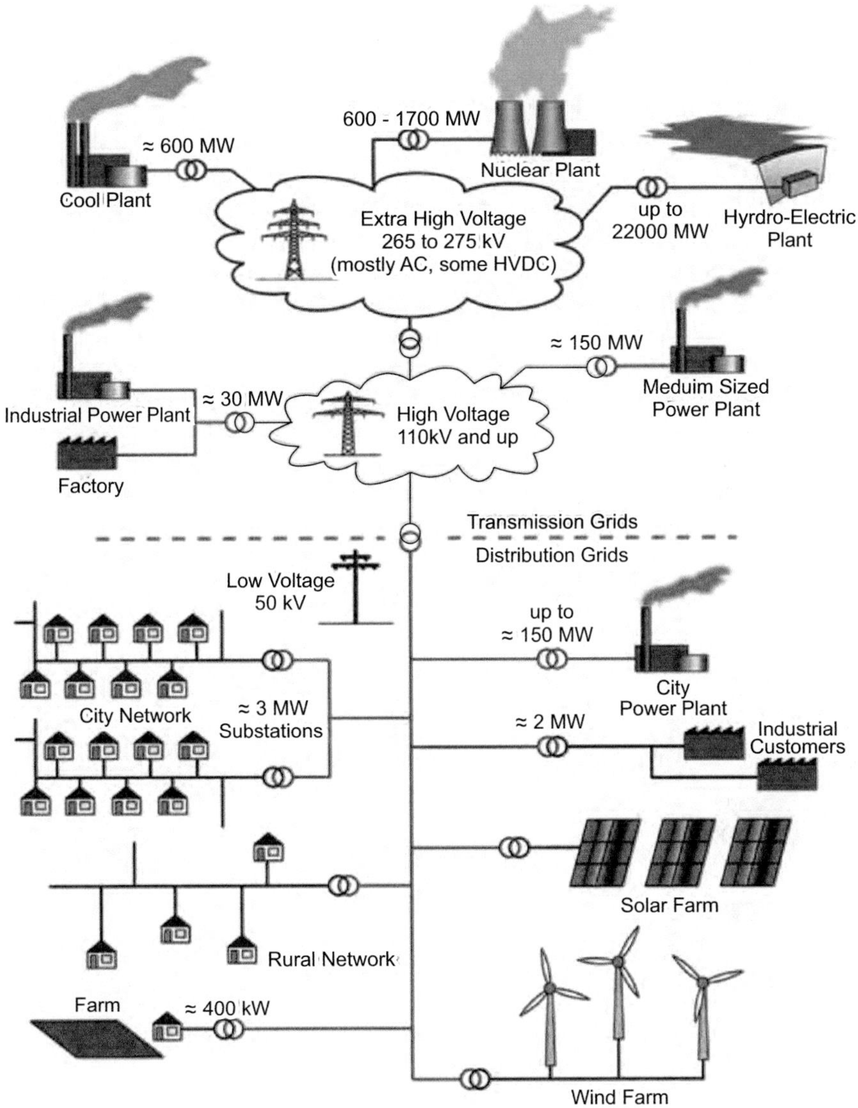

Fig. 1.4 The electricity grid in a nutshell. [15]

infrastructure, with problems particularly inherent to the interconnection of networks. Many of these networks are, due to their design, limited in their capacity to receive decentralized energy supply.

Smart grids are systems that use information and communication technologies to optimize the operation of the electrical system, exchanging data and electricity between the various actors in the network. They differ from traditional electrical grids, which, as previously described, are primarily unidirectional systems that transport electricity from production sites to end consumers, with limited communication or control between them.

Smart grids offer numerous advantages, such as improving the reliability, security, and efficiency of the network, facilitating the integration of renewable energy sources and power storage systems, and involving consumers in the management of their electricity consumption and production. They rely on several technologies, such as: smart meters, which measure and transmit real-time consumer electricity consumption and production; smart distribution networks, which control and optimize the flow of electricity on low and medium voltage lines; smart transport networks, which monitor and regulate the flow of electricity on high voltage lines; and microgrids, which are autonomous networks that locally manage the production and consumption of electricity from onsite energy sources.

The transformation from a traditional electrical grid to a smart grid involves several steps, requiring coordination between the various actors in the electrical system, as well as an adaptation of regulatory, economic, and social frameworks. The goals of this transformation: are ensuring a reliable and resilient power supply, maintaining a competitive cost of electricity, and obtaining environmental benefits.

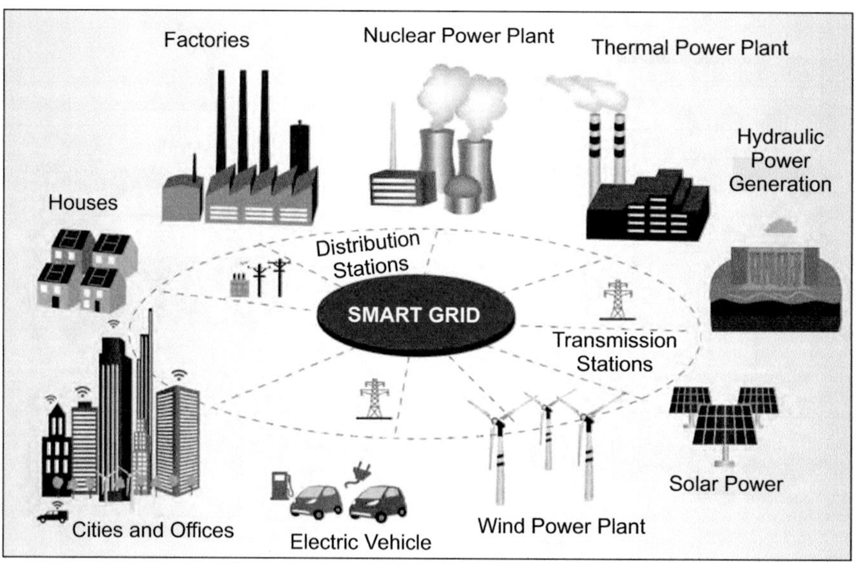

Fig. 1.5 The smart grid in a nutshell. [16]

Figure 1.5 shows the smart grid, which is an advanced version of the traditional electricity grid. It uses modern technology to improve efficiency, reliability, and sustainability. The smart grid uses digital technology to manage the flow of electricity from all sources to meet end-user demand. It uses renewable energy, makes the grid safer, and gives utilities and consumers real-time data. The smart grid allows utilities and customers to talk to each other. This helps manage demand

and respond to outages faster. Advanced Metering Infrastructure (AMI) includes smart meters that provide real-time data on electricity usage. This helps consumers manage their energy consumption and allows utilities to detect and address issues promptly. The smart grid can easily use renewable energy like solar and wind, which reduces reliance on fossil fuel supplies, helps improve air quality, and mitigates the impacts of climate change. The smart grid also uses energy storage to preserve energy when generation exceeds load and discharge it when load exceeds generation. Many traditional electricity grids use one-way communication with analog meters that in some cases are even still read manually. Consequently, consumers in the traditional grid often have limited visibility into their energy usage. The smart grid allows consumers to monitor and manage their energy consumption in real-time.

In broad terms, the differences between a traditional electrical grid and a smart grid are as follows:

- A traditional grid is unidirectional, while a smart grid is bidirectional.
- A traditional grid is passive, while a smart grid is active.
- A traditional grid is rigid, while a smart grid is flexible.
- A traditional grid is opaque, while a smart grid is transparent.

A traditional grid only flows in one direction—from the power plant to the consumer. A smart grid is different. It works in both directions, allowing electricity and information to flow in both directions. This better manages and integrates renewable energy sources. Traditional grids are passive, simply delivering electricity without much interaction, while smart grids are active, constantly monitoring and responding to changes in demand and supply. This makes traditional grids rigid and less adaptable to changes, whereas smart grids are flexible, able to adjust to varying conditions and incorporate new technologies. Lastly, traditional grids are opaque, providing limited information to consumers and operators, but smart grids are transparent, offering real-time data and insights to improve efficiency and reliability.

Challenges of Traditional Grids and Technological Innovations Leading to Microgrids

The traditional electrical grid is becoming increasingly vulnerable. One of the main problems is that the infrastructure is aging, inflexible and not designed to meet the needs of a growing population and the increasing diversity and sizes of electrical loads from newly electrified systems (e.g., heating and transport) and novel technologies (e.g., AI and the data centers needed to power it). This old infrastructure fails more often than it previously did, causing power outages, less reliable service and greater fluctuations in power quality. Also, traditional grids are centralized, so they can be disrupted by natural disasters, cyber-attacks, and other catastrophic events. If one part fails, it can cause a blackout that affects millions. Long-distance transmission lines also lose energy over distance and make the system less efficient. These lines can also be easily damaged or rendered inoperable by weather, such as hurricanes, wildfires, and ice storms, which are becoming more common and severe due to climate change. The operation of many traditional grids also is detrimental to the environment. Many still predominantly rely on the combustion of fossil fuels for power generation, which emits large quantities of greenhouse gases and air pollution. This dependency amplifies the effects of climate change and also harms people living near thermal power plants. We need to change how we make, send, and use energy. These imperatives are leading us to create more reliable and friendly solutions like microgrids.

The commercialization and proliferation of new technologies is helping facilitate the deployment of microgrids. Advances in and lower costs related to renewable energy, especially solar and wind power, have been very important. Solar panels are now more efficient and affordable, so small-scale installations can pay for themselves much more quickly than before. Wind turbines are now more efficient and cheaper, so they can be used in microgrids where average winds speeds make them practical. Lithium-ion batteries have also improved. These batteries store surplus energy production and release it when it is needed, ensuring a reliable power supply. Smart grid technology is also important for microgrids. Smart meters, sensors, and control systems let you monitor and manage energy production, distribution, and use in real time. These technologies let microgrids work on their own, use energy wisely, and react quickly to changes in demand or supply. AI and machine learning make microgrids more efficient and reliable. They predict energy usage, optimize energy storage, and identify potential issues before they cause failures. Blockchain technology has also made it easier for people in microgrids to trade energy with each other. Blockchain technology is a digital record of transactions that is spread across many computers. This record is secure and transparent. Each transaction is added to a "block" and linked to the previous one, forming a "chain". This makes it very useful for trading energy within microgrids because it ensures that all transactions are secure and transparent and cannot be altered. Participants can trade energy directly with each other without the need for a central authority, enhancing the efficiency and reliability of the energy market. This can lead to a more decentralized and democratic energy market. These new technologies can make microgrids a good, reliable choice for sustainable energy.

On the Composition of Future Grids

Now that we have defined what the past and present look like in terms of electrical networks, we can dive into what composes them in more detail. An electrical network is composed of several elements, such as production plants or sources, electrical lines, electrical substations, transformers, and power consumers. These elements must be managed optimally to ensure the safety, quality, and efficiency of the electrical service. To improve its operation and efficiency, a smart grid will add additional elements to the traditional electrical network, such as sensors, actuators, communication devices, analysis systems, and smart meters.

A Fictitious Example

Imagine a small town named Greenville in a valley. Greenville has adopted a microgrid to manage its electricity more efficiently and sustainably. Let's look at how Greenville's microgrid works. Greenville's microgrid has sensors spread throughout the town. These devices are on power lines and in homes and businesses. They measure the flow of electricity, tracking how much power is used, the current quality, and any spikes or drops. A sensor notices a drop in voltage on Main Street. The control center is immediately informed. Communication devices send the data from the sensors to the control system using radio waves and fiber optic cables. The control center gets the alert about the drop in voltage and starts to investigate it.

The control center has advanced analysis systems. These systems use AI and machine learning to process data. They find that a tree branch has fallen on a power line, causing the drop in voltage. The system sends a command to the actuators. Actuators are the muscle of the microgrid. They adjust the voltage and reroute the electricity to keep the supply stable. The actuators also tell a maintenance team where the problem is.

Greenville residents also use smart meters that show how much electricity is being used at any given moment. On a hot day, the smart meters show a spike in electricity use as air conditioners run. The microgrid's system predicts this increased demand and adjusts the power distribution. This ensures everyone stays cool without overloading the system. Greenville's microgrid uses a network of sensors, communication devices, analysis systems, actuators, and smart meters to operate smoothly and efficiently. The town thereby can maintain a reliable and sustainable power supply, with residents able to conserve energy as a bonus. This story illustrates how a microgrid links various technical elements and technologies to create a resilient and intelligent electrical network.

Solar Microgrid Overview

A solar microgrid is a small energy system that uses solar panels to make electricity. A solar microgrid has four main parts: solar panels, an inverter, a battery, and a control system. The PV panels capture sunlight and turn it into electricity. The inverter changes the DC electricity into AC electricity, which can be used in homes and businesses. The battery stores excess electricity made during the day, which can be used at night or in low sunlight. The control system manages the electricity, ensuring a reliable supply. Solar microgrids can work without the main power grid, providing electricity in remote areas or during power outages. They are also environmentally friendly. Figure 1.6 presents a conceptual overview of data and energy flow in a smart microgrid, which can include both local and main grid connectivity. Some microgrids can work without the main power grid, providing electricity in remote areas or during power outages.

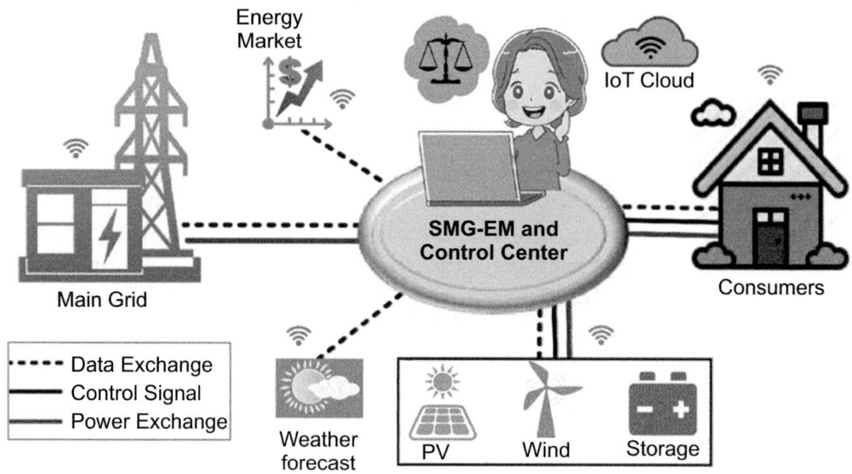

Fig. 1.6 Conceptual overview of data and energy flow in a smart microgrid system [17].

The Future of Sustainable Mobility: Embracing Smart Electric Vehicles and Renewable Energy

Another driver of microgrid proliferation is the increasing adoption of electric transport, especially personal transport. Electric vehicles (EVs) are perfect for integrating into a microgrid powered by renewable energy sources like solar panels paired with energy storage solutions. The growing adoption of EVs and development of related battery technologies will also offer potential microgrid developers more options in terms of power storage types and systems and reduce the strains on supply chains related to the currently predominant energy storage technology in use for EVs and stationary systems—lithium-ion batteries.

These batteries rely on metals like lithium, cobalt, nickel, and some rare earth elements. As demand increases, we will need more of these metals. Mining and extracting them can harm the environment and create reliance on one or a few global suppliers, which can lead to geopolitical tension and increase supply chain risk. To address these challenges, we need to find ways to use less of these materials by developing batteries that rely on different raw materials. Governments and industries in the US, Europe, and China are already working to develop alternatives to the lithium-ion battery or more efficient versions of it. Governments and industries in the US, Europe, and China are already working to develop new types of batteries that are better than lithium-ion batteries. Some promising options include sodium-ion batteries, which are more affordable and better for the environment, and solid-state batteries, which are safer and can store more energy. They are also looking into lithium-sulfur batteries and graphene-enhanced lithium-ion batteries to improve how much energy they can store and how fast they can charge. These innovations are meant to deal with the shortcomings of current lithium-ion technology and support the growing demand for efficient and sustainable energy solutions.

To better understand the infrastructure needs of the energy transition, we turn to a data analysis based on recent reports [18]. Global battery storage power capacity stood at approximately 230 gigawatts (GW) in 2022. Under current policy trajectories, Bloomberg NEF (via the Sustainable Energy in America Factbook) projects this will nearly triple to 650 GW by 2030. More ambitious climate pathways require even steeper growth: the International Energy Agency's Net Zero by 2050 roadmap indicates battery storage must hit 778 GW by 2030 and rise dramatically to 3,860 GW by 2050 to support deep decarbonization and integration of variable renewables.

To accelerate progress, G7 nations have committed to a global energy storage goal of 1,500 GW by 2030—about six times the 2022 level—aligning with the COP28/29 pledge to triple renewable capacity and expand storage infrastructure.

This widespread commitment underscores that large-scale energy storage is now fundamental for enhancing grid flexibility, supporting rising electricity demand, and advancing resilient carbon-neutral power systems.

Social Aspects of the Energy Community

Why not stop depending on the electrical grid and move to a fully autonomous microgrid or partial autonomy via an interconnected, islandable microgrid? There are many reasons to go off-grid, including wanting to be more independent, save money, and reduce one's carbon footprint. As electricity demand increases, it may be good to rely less on centralized power grids that are struggling to keep up with the rapidly increasing demands on their supply. As battery storage and solar power become cheaper and utility rates rise, it is becoming more attractive for people to become prosumers by installing microgrids. "Prosumer" is a combination of the words "producer" and "consumer". It reflects the dual role people can play in generating and consuming energy. As a prosumer, you are both a consumer and a producer of energy. This shift towards prosumers can help create energy communities, where people work together to manage and share energy resources, much as many communities now work together to maintain shared gardens. People of all ages work together to take care of the energy community, much as they do when caring for a garden. This can be a model how microgrids can help communities and people to gain greater or total control of their energy resources.

This collaborative endeavor exemplifies the essence of microgrids, a revolutionary approach not only to energy but also to fostering community and personal empowerment. Many microgrids are designed and implemented to be "energy islands" that will continue operating if disconnected from the main grid. This is called "islanding" mode. The proliferation of islandable microgrids is changing communities by allowing them to be more autonomous. This evolving trend shows how communities are working together to solve energy challenges and become more resilient and powerful (pun again intended).

These advantages inherent to microgrids make it possible for this story to be rich and deep. One may easily imagine sustainable communities being built in part on the premise of shared social responsibility improving living standards and enhancing security. These microgrids can both provide electricity to homes and help communities shape their energy future. As we move from centralized power to a more diverse system, we must think about the challenges of our modern world, why we are making these changes and how this can help us decide whether to use a microgrid in our community. The journey towards energy autonomy is not just about technology; it is also about the economy and society.

With microgrids, people can make energy solutions that fit their needs, like saving money, maximizing resilience, protecting the environment, mitigating the effects of climate change, reducing reliance on outside companies for energy security, or a combination of these. These customization possibilities are similar to designing a garden to provide specific benefits. Community workshops organized by local government, related businesses and non-profit organizations can help people understand these topics and how microgrids can benefit communities and companies. Implementing microgrids can also teach about renewable energy, conservation, and the environment. Simpler microgrids typically used for residential electrical loads are relatively easy to install and manage and let people take greater control over their energy management more quickly and effectively. Like tending to a garden, it engenders in people a sense of ownership and responsibility.

Important Factors for the Public to Increase the Sale of Energy Storage Systems

This section explains why people want to be energy independent and what they think about it. Understanding these factors helps to understand why people and communities are driven to invest in energy storage solutions. These solutions are integral to the functionality and efficiency of microgrids. We can develop comprehensive strategies that promote the widespread adoption of energy storage systems by addressing financial incentives, brand trust, technological diversity, and customer education. This will ultimately support the transition to energy autonomy.

The factors deemed important or influential in increasing the sales of energy storage systems, based on the opinions of photovoltaic system installers in Italy in 2019, include tax credits and regional incentives, which were rated as very important by a significant majority of respondents [19]. Other notable factors include the trust in and reputation of brands and products, the multiplicity of available technologies and solutions, and the service assistance offered by vendors. Additionally, the knowledge of the topic among customers and the faster return on investments were also highlighted as influential. These insights underscore the multifaceted considerations that impact the adoption and sales of energy storage systems, emphasizing the need for comprehensive strategies that address financial incentives, brand trust, technological diversity, and customer education.

If you want to help communities become more independent, save money, and mitigate the impacts of climate change, you can combine different energy sources and storage solutions into your microgrid design. This helps the country become more energy independent, use energy more efficiently, and save money. It also helps us deal with rapidly rising utility costs, become more energy resilient, and improve air quality. A well-designed microgrid can be a sustainable and resilient energy solution for everyone.

Solar power is common to find as part of a microgrid, but other power sources are needed for a microgrid to be islandable and to establish more than very limited autonomy. Investing in solar panels can result in long-term cost savings and increased energy independence. However, solar is by nature an intermittent power source dependent on the availability of sunlight, which varies based on the weather and seasons. Non-intermittent power sources operate independently of solar or wind conditions (e.g., hydroelectric, nuclear, and fueled generation).

To be islandable, microgrids must incorporate power storage (usually batteries), as well as non-intermittent power sources (typically generators fueled with natural gas or diesel) to ensure power availability in the event of a grid outage. For larger microgrids, pumped hydro storage can also store energy made during the day for use at night or on days with little sunlight. Forms of power storage based on thermal batteries and batteries with much longer operational lives than lithium-ion are currently in development. Furthermore, for those focused on the environmental and climate change mitigation benefits microgrids can yield, it is possible, though typically more expensive, to fuel non-intermittent sources with renewable methane, biomass or biodiesel. While more expensive, it is typically easier to do this with locally sourced renewable fuels (potentially available from anaerobic digesters on municipal wastewater streams or agricultural waste sources, biomass gasification using pyrolysis or thermolysis, waste oils from restaurants, homes, etc.), which increases the end user's power autonomy and reduces reliance on refined petroleum products or pipeline natural gas supplied large corporations.

A microgrid will be more self-sufficient, resilient, and flexible when all these elements are combined. Most islandable microgrid operators will probably opt to remain interconnected with the grid, at a minimum to have the grid as a backup source of power if some part of the microgrid fails. It also makes sales of power to the grid via net metering or renewable energy credits, as well as integration into a virtual power plant (discussed later), possible. Considering these factors, the payback period and overall economic viability of the project depends on various factors, such as location, available incentives, system configuration and complexity and local electricity rates. It is recommended to consider these factors carefully before making any decisions.

The main drawbacks for residential solar power adoption in the United States in 2022 were identified in a study by Forbes [20]. The most significant barrier, cited by 36% of respondents, is the upfront cost of installation. This is followed by a variety of other concerns, including 10% of respondents who reported having no hesitations about adopting solar power. Other notable drawbacks include financial priorities, with 9% of respondents indicating they have other life events to pay for and 8% planning to move before they would see savings from solar power. Maintenance and reliability of electricity are each cited by 8% and 7% of respondents, respectively. Additionally, 7% of respondents are concerned about the monthly costs associated with solar power, while 5% are worried about the annual days of sun in their region. Aesthetic concerns are minimal, with only 1% of respondents indicating this as a drawback. The study underscores the diverse range of factors that influence the decision to adopt residential solar power, highlighting the need for addressing financial and practical concerns to increase adoption rates.

Financial Incentives

Many countries offer financial incentives to install renewable power that can supply microgrids. Nevertheless, many microgrids remain powered, at least in part, by traditional generation, i.e., turbines and generators combusting natural gas or diesel. This is mainly due to cost, with traditional power sources currently offering competitive $/kWh cost of electricity, but also with familiarity. To transition reliance from non-renewable to primarily renewable energy generation to maximize autonomy and environmental benefits, microgrid developers wishing to use renewable energies need to seek out and leverage all available financial incentives. They will reduce the upfront and lifecycle costs of installing and operating a microgrid and its related technologies.

For example, in the US, the federal government currently offers a Solar Investment Tax Credit (ITC) and Residential Clean Energy Tax Credit, which is a percentage of the installation cost deducted from your federal taxes. As of 2022, these credits are 30%. It is important to note that these incentives may not be permanent and may be revised or canceled in the future. Additionally, some states and local governments offer further incentives, rebates, or tax credits for solar installations. The problem with this type of regulation is that it is often different from one region to the other. In Canada, available energy incentives are different in every province.

To understand how recent geopolitical shocks have reshaped Europe's energy landscape, we turn to data from the International Energy Agency and the European Commission [21].

Geopolitical tensions have had a profound impact on European electricity markets, fundamentally altering price dynamics and emphasizing the importance of robust energy security strategies. In 2020, the average wholesale electricity price in Europe was around €35 per megawatt-hour (MWh). By December 2021—prior to Russia's invasion of Ukraine—this price had surged to nearly €250/MWh, representing an increase of over 400%.

Following Russia's invasion of Ukraine in February 2022, the IEA described the event as "the first truly global energy crisis". In March 2022, average electricity prices spiked above €500/MWh. This escalation reflected Europe's heavy dependence on Russian gas: in 2021, Russia supplied 45% of the EU's gas imports and accounted for 23% of total energy imports in the Euro area.

Because European electricity prices are typically set through marginal gas pricing, any spike in gas costs directly pushes up power prices across the continent. In response, the European Commission implemented the REPowerEU Plan in May 2022. By 2024, Russian gas imports into the EU had fallen from 150 billion cubic meters to 52 billion cubic meters, reducing Russia's share from 45% to 19%.

These developments highlight the critical need for diversified energy sources, increased storage capacity, accelerated renewable deployment, and strengthened international cooperation to build long-term energy resilience, and reduce vulnerability to geopolitical shocks.

Integrating renewable energy into future grids is a complex challenge. It requires a comprehensive approach to ensure reliability, efficiency, and sustainability. Different renewable energy sources have different characteristics and ways of working and must be made to work together in the grid. Solar and wind energy are intermittent and unpredictable, so we need to use advanced forecasting and flexible grid management to balance supply and demand. Hydroelectric and geothermal power can complement the variability of solar and wind and make the grid more stable. The integration process also involves upgrading existing infrastructure to accommodate distributed generation, storing excess energy, and developing strong connections to transfer energy across regions. Mixing different renewable sources thus makes future grids more resilient and sustainable. It also reduces dependence on fossil fuels and helps mitigate the effects of climate change. Smart grids play a critical role in managing and optimizing energy distribution in an increasingly complex and decentralized energy landscape. These advanced grid systems use digital technology, real-time data analytics, and automated control mechanisms to improve the efficiency, reliability, and flexibility of energy distribution. Smart grids enable the integration of renewable energy sources by providing the infrastructure necessary to monitor and manage energy flows in real time. They facilitate demand response programs that allow consumers to adjust their energy use based on grid conditions and price signals, thereby reducing peak demand and increasing grid stability. In addition, smart grids support the deployment of distributed energy resources (DERs), such as rooftop solar panels and home battery systems, by enabling bidirectional energy flows and ensuring that excess energy can be stored or returned to the grid. The deployment of smart meters and advanced metering infrastructure (AMI) provides consumers with detailed insights into their energy consumption patterns, promoting energy efficiency and conservation. By fostering greater interaction between utilities and consumers, smart grids create a more dynamic and responsive energy ecosystem capable of meeting the evolving demands of the 21st century.

To conclude this section, the emergence of the prosumer is a key factor in the microgrid revolution. Whereas traditional energy systems were based on centralized production and passive consumption, microgrids enable individuals to become prosumers, i.e., both producers and consumers of energy. This paradigm shift has far-reaching implications for the way we think about and manage energy.

Here are a few concrete examples of prosumers and their impact on the energy market:

- A homeowner installs solar panels on his roof and a battery to store excess energy. He can then supply his house with clean energy and even sell the surplus to the electricity grid.
- A community comes together to create a microgrid powered by renewable energy sources. Community members share the energy produced, reducing their dependence on the main grid, and cutting their electricity bills.
- A company invests in a cogeneration system that produces both electricity and heat from a single fuel source. This enables the company to cut its energy costs and reduce its environmental impact.

Challenges and Solutions for Prosumers

Despite the many advantages of microgrids, prosumers face several challenges:

- **High Initial Costs:** The initial investment to install solar panels, batteries, and other equipment can be significant.
- **Technical Complexity:** Managing a microgrid can be complex and require specific technical skills.
- **Regulatory Framework:** Regulations on self-generation and selling energy to the grid can be unclear or restrictive.

Many potential solutions to these problems are also emerging:

- **Financial Incentives:** Grants, tax credits, and low-interest loans can help reduce initial costs.
- **Turnkey Solutions:** Specialist companies offer turnkey microgrid solutions, simplifying installation and management.
- **Evolving Regulatory Framework:** Governments are working to adapt regulations to encourage self-generation and prosumer participation in the energy market.

This brings us to the next section. The transition to a more decentralized and participatory energy model, in which prosumers play a key role, also requires technological innovation and effective data management.

Microgrids: Revolutionizing Energy Distribution

In the ever-evolving landscape of energy distribution, microgrids have emerged as offering resilience, sustainability, and efficiency. Picture a world where traditional power grids are no longer the sole solution, where localized energy generation and management take center stage and offer a myriad of benefits to communities and industries alike.

The journey into understanding microgrids begins with acknowledging the vulnerabilities of centralized power systems. Massive blackouts, like those witnessed in Texas and California in 2021, serve as stark reminders of the fragility of traditional grids in the face of extreme weather events. These incidents underscore the urgent need for alternative energy solutions that can withstand such challenges.

Increasingly frequent and more powerful hurricanes and winter storms damage power grids and greatly reduce their reliability. For example, Hurricane Ida damaged Louisiana's electric grid and knocked out power to over 1 million customers. It took weeks to fully restore electricity. The grid was particularly vulnerable because of old infrastructure and damage from other hurricanes. The storm damaged 30,000+ utility poles, 35,000+ spans of wire, and 5,600 transformers [22]. Storms are damaging power grids in Europe. Storm Arwen, one of the most powerful winter storms in recent years, hit Scotland and northern England in late November 2021. Strong winds and fallen trees damaged energy infrastructure, leaving over a million homes without power. Some areas were without power for over a week. The grid was made more vulnerable by old infrastructure and previous damage. Storm Arwen damaged utility poles, wire, and transformers. This shows the need to modernize the grid to withstand extreme weather and ensure a more resilient power supply [23].

Decentralized networks bring power generation and consumption closer together, reducing reliance on distant power plants and long-distance transmission lines. At the heart of most microgrids lie renewable energy sources like solar panels and wind turbines paired with battery power storage, equipped with sensors and controllers to optimize their utilization. These systems not only provide emission-free energy but also offer greater reliability through real-time monitoring and management.

But the transformation to microgrids is not just about technology; it is also a strategic shift driven by diverse stakeholders, including business executives that recognize the transformative potential of this approach. It also involves reshaping traditional utility models, embracing open data management, and fostering collaboration among experts to drive innovation forward.

One of the key advantages of microgrids lies in their ability to adapt to changing energy demands and weather patterns. Unlike traditional grids, which do possess some degree of flexibility based on circumstances, microgrids are designed to be even more flexible and resilient. They can better withstand weather extremes and accommodate rapid load rebalancing driven by a variety of residential and commercial end users. This enhanced adaptability makes microgrids particularly valuable in scenarios where rapid response to fluctuating conditions is crucial, such as during natural disasters or in remote locations with variable energy needs from different users.

Moreover, microgrids empower communities by democratizing energy access, offering opportunities for job creation and economic development. They address energy poverty by providing affordable and reliable electricity to underserved areas, while also promoting sustainability and environmental stewardship.

Yet challenges remain. From cybersecurity concerns to the need for skilled engineers for installation and maintenance, the path to widespread microgrid adoption is not without obstacles. However, with nearly 8 in 10 Americans supporting their use once informed [24], there is a growing recognition of microgrids' potential to revolutionize the energy landscape.

In essence, microgrids represent a paradigm shift—a departure from the centralized, one-size-fits-all approach of traditional grids towards a decentralized, community-driven model. They embody resilience, sustainability, and empowerment, offering a glimpse into a future where energy is not just a commodity but a catalyst for positive change. As we navigate the complexities of our energy transformation, microgrids stand as beacons of hope, illuminating a path towards a brighter, more sustainable future for all.

Social and Community Impacts of Microgrids

Microgrids help communities and businesses save money. Microgrids can cut energy costs for communities by generating power locally, avoiding the high costs of long-distance electricity. This also means less energy is lost in transmission, so what is generated can be used effectively. Microgrids also provide a stable energy supply, protecting communities from price volatility. Using solar and wind energy helps communities use less fossil fuel, making them ultimately more independent. Microgrids can also create jobs and help the economy grow. These jobs include installing and maintaining renewable energy systems and managing and operating the microgrid. Microgrids also make businesses more competitive by giving them a comparatively more secure and reliable energy supply.

Power outages are expensive for everyone but especially for businesses and governments. Microgrids ensure power keeps flowing when the grid fails. This is especially important for industries that need a reliable energy supply, like manufacturing, healthcare, national defense, and data centers. In some jurisdictions, businesses can also save money by making their own power and selling extra energy back to the grid. This creates new revenue streams and allows the business to tout itself as helping to mitigate the impacts of climate change, instead of exacerbating them. Businesses that invest in microgrids can also get government incentives and tax credits for using renewable energy.

The economy affects the decision to pursue energy independence. Utility-supplied power is often cheaper in the short term, making most microgrids initially an expensive proposition. With careful planning, the right incentives, and advanced technology, a microgrid can often be economically justified.

Microgrids change how we make, share, and use energy. It is not just a technical challenge. It's also a response to global issues like rising energy costs, increasing damage from climate change, and the need for reliable, clean energy. Microgrids give people and communities more control over their energy supply, which helps make energy more accessible, especially in remote or underserved areas.

Conclusion: Weighing the Case for Energy Autonomy

It is clear that becoming energy independent is not just a technical challenge. It is a multifaceted challenge that includes economic, environmental and, in some cases, societal aspects. Humanity's journey with electricity has gone from the mystical to the practical, witnessing along the way many amazing moments: first harnessing the power of the gods; creating the first battery; power generation using hydroelectric and thermal power plants; the establishment of centralised grids; the development of nuclear, solar, wind, and geothermal power sources, scalable power storage systems; and now the deployment of islandable, scalable microgrids that brings all of these together to be available to individual households, businesses, communities, and other institutions.

In the 21st century, it is clear we need another way of producing and distributing energy. Ageing infrastructure and the steadily increasing risk of more powerful natural disasters and cyber attacks have shown that we can't solely rely on an interconnected grid. Also, the environmental contamination caused by large-scale fossil fuel use and its amplification of the effects of climate change have made us realize that we need to integrate low-carbon and zero emission.

Microgrids offer an attractive option for addressing many of these challenges. They can integrate a variety of energy sources, including solar, wind, hydroelectric, power storage, and fueled power generation. Given power generation is almost always colocated or located very close to the end users of the power, microgrids are not exposed to the same transmission failure risk one finds with traditional interconnected grids. Consequently, microgrids offer a more sustainable, resilient, and reliable alternative to many traditional energy systems.

The societal impacts of microgrids are also important to consider. Microgrids can give people more control over their energy supply and reduce reliance on big energy companies. This can mean better and wider access to reliable energy, especially in remote areas and communities that currently have below-average power availability. New economies, supply chains and employment opportunities can arise as a result too which by necessity will be very locally oriented. This can create sustainable and durable economic prosperity in the places where microgrids are deployed and particularly where they can be aggregated into virtual power plants.

As we move forward, it is important to understand that the journey towards energy autonomy does not need to be only about technology and economics. It is about people working together to create energy systems that are sustainable, resilient, and autonomous. By using microgrids, individuals, businesses, and communities can help foster a more secure and prosperous future.

References

1. https://www.history.com/topics/inventions/telegraph.
2. https://www.loc.gov/collections/samuel-morse-papers/articles-and-essays/invention-of-the-telegraph/
3. https://energy-cities.eu/the-evolution-of-public-lighting-from-torches-to-smart-services/.
4. https://www.energy.gov/articles/history-light-bulb.
5. https://fr.wikipedia.org/wiki/La_F%C3%A9e_%C3%89lectricit%C3%A9).
6. https://19january2021snapshot.epa.gov/greenpower/us-electricity-grid-markets_.html.

7. https://www.epa.gov/green-power-markets/us-electricity-grid-markets.
8. https://www.entsoe.eu/data/map/.
9. Ember. (2024). *Global Electricity Review 2024*. Retrieved June 8, 2025, from https://ember-climate.org/insights/research/global-electricity-review-2024/.
International Energy Agency. (2024). *Electricity 2024: Analysis and Forecast to 2026. IEA*. Retrieved June 8, 2025, from https://www.iea.org/reports/electricity-2024.
10. https://www.npr.org/2022/10/05/1126900340/florida-community-designed-weather-hurricane-ian-babcock-ranch-solar.
11. https://www.taylorfarms.com/newsroom/taylor-farms-taps-bloom-energy-ameresco-and-concept-clean-energy/.
12. As with the example of the Florida courts, these laws may end up being in regulations not directly related to the energy sector but instead in the realm of property law or safety standards. Your jurisdiction may have no such laws, or its laws may prohibit disconnection without requiring substantial proof that residences or other affected structures will remain safely habitable and usable.
13. https://www.microgridknowledge.com/editors-choice/article/11436878/major-north-american-food-producer-to-exit-the-california-grid.
14. The Earth and I. (2024). *Global Electricity Report 2024*. Retrieved June 8, 2025, from https://www.theearthandi.org/post/global-electricity-report-2024.
Ember. (2024). *Global Electricity Review 2024*. Retrieved June 8, 2025, from https://ember-climate.org/app/uploads/2024/05/Report-Global-Electricity-Review-2024.pdf
Cleveland, C. and Clifford, H. (2023). *World Electricity Generation since 1900. Visualizing Energy*. Boston University Institute for Global Sustainability. Retrieved June 8, 2025, from https://visualizingenergy.org/world-electricity-generation-since-1900/.
15. The electricity grid in a nutshell (*Source:* Wikipedia "Electric grid". https://commons.wikimedia.org/wiki/File:Electricity_Grid_Schematic_English.svg#/media/File:Electricity_Grid_Schematic_English.svg.)
16. The smart grid in a nutshell. (*Source:* https://www.bfe.admin.ch/bfe/en/home/supply/electricity-supply/electricity-networks/smart-grids.html.)
17. Saeed, M.A., Sedhom, B.E., Elbaghdadi, A.S., Fadl, A.W., Abdelwahab, M.S., Elden, M.H., ... and Eladl, A.A. (2023). Practical prototype for energy management system in smart microgrid considering uncertainties and energy theft. *Scientific Reports, 13*(1), 20812. (*Source*: https://www.nature.com/articles/s41598-023-48011-w?fromPaywallRec=false.)
18. Business Council for Sustainable Energy & Bloomberg NEF. (2025). 2025 *Sustainable Energy in America Factbook* [PDF]. Retrieved June 8, 2025, from https://bcse.org/wp-content/uploads/2025/02/2025-Sustainable-Energy-in-America-Factbook.pdf.
International Energy Agency. (2023). *Net Zero by 2050: A Roadmap for the Global Energy Sector*. Retrieved June 8, 2025, from https://www.iea.org/reports/net-zero-by-2050.
COP29 Azerbaijan. (2024). *Global Energy Storage and Grids Pledge: Background Information*. Retrieved June 8, 2025, from https://cop29.az/en/pages/global-energy-storage-and-grids-pledge-background-information.
19. https://www.solareb2b.it/newsletter/SolareB2B-mag2019-hd.pdf.
20. Forbes. https://www.forbes.com/home-improvement/solar/solar-energy-pros-and-cons/.
21. International Energy Agency. (2024). *Russia's War on Ukraine: Energy Implications and Responses*. Retrieved June 8, 2025, from https://www.iea.org/topics/russias-war-on-ukraine.
European Commission. (2024). *REPowerEU: Three Years On*. Retrieved June 8, 2025, from https://energy.ec.europa.eu/topics/markets-and-consumers/actions-and-measures-energy-prices/repowereu-3-years_en.
22. https://www.politico.com/news/2021/08/30/hurricane-ida-new-orleans-louisiana-rescues-507504.

23. https://theconversation.com/how-extreme-weather-threatens-to-bring-down-uks-power-lines-and-halt-supply-to-homes-203193.
24. Jamil, N., Qassim, Q.S., Bohani, F.A., Mansor, M. and Ramachandaramurthy, V.K. (2021). Cybersecurity of microgrid: State-of-the-art review and possible directions of future research. *Applied Sciences*, *11*(21), 9812. https://www.mdpi.com/2076-3417/11/21/9812.

2 The What: Building Blocks of Energy Independence

Chapter 2 provides a thorough examination of the technologies and components that enable microgrids to be viable, sustainable, and scalable. Readers are introduced to the essential building blocks of energy independence, including renewable energy sources (e.g., solar, wind, and hydro), energy storage systems (e.g., lithium-ion batteries), smart grid technologies, and advanced energy management systems. It contrasts the rigid, unidirectional structure of traditional grids with the flexibility and intelligence of smart grids, which enable two-way communication and decentralized power control. Clear illustrations and analogies demystify technical concepts such as voltage regulation, power quality, and load balancing. Through case studies and hypothetical examples, such as the fictional town of Greenville, the chapter illustrates how sensors, actuators, and AI-enabled systems interact within a functioning microgrid. The chapter emphasizes the integration of electric vehicles, noting their dual role in transportation and as mobile energy storage. It also explores evolving battery technologies, including promising alternatives to lithium-ion batteries. The chapter addresses environmental impacts, infrastructure challenges, and material sourcing for energy storage, as well as emerging global targets for storage capacity set by the G7 and IEA. Throughout, the authors advocate for microgrids as a solution to the technical shortcomings of legacy grids and as a strategic pathway to achieving carbon neutrality, energy resilience, and democratized power systems. This chapter provides readers with the technical knowledge necessary to understand and evaluate energy autonomy solutions in their own contexts.

So you have decided to install a microgrid or join with others to do so. In Chapter 1, we went through the historical context, as well as some key considerations and challenges to consider before embarking on this journey, and decided "yes, let's do this". In this chapter, we dive into the "what". Namely, that includes "what" kind of microgrid you will install. "What" you will use it to power, "what" kind of system you design, and "what" equipment you will need to purchase.

We discuss in this chapter the various technical aspects of handling electricity, such as production, storage, and transportation via an electrical grid. We also talk about electric consumption applications like heating, cooling, lighting, residential and industrial appliances, and systems and mobility. Further, we will detail network-connected systems (on-grid), network-disconnected systems (off-grid), microgrids, and smart grids. Finally, we discuss their energy generation systems, whether they are powered by fossil fuels or renewable energies like the sun for

DOI: 10.1201/9781032689845-2

solar panels, wind for wind turbines, water for hydroelectricity, biomass and waste streams for biofuels and renewable gas, and subsurface heat for geothermal. Additionally, we will discuss energy conversion via inverters or converters, direct and alternating current and voltage, and power, and capacity.

But where to start…? We assume the reader is already familiar with the basic physical aspects of electricity and energy in general and so will commence by introducing the concept of the electrical grid. This basis will then allow us to elaborate on the weaknesses and limits of the current electrical networks, which we will call the traditional electrical grid, and to examine current and future techniques that provide the basis for defining what a smart grid is and can be.

Energy Production

Let's talk first in more depth about energy production. There are three main sources of power generation currently for grids that we will detail below. These are fossil fuel energies, nuclear and renewable energy. According to the International Energy Agency, global electricity generation in 2023 was distributed as follows:

- Fossil fuels: 59%
- Hydropower: 15%
- Nuclear: 9%
- Solar and wind: 13%
- Bioenergy and geothermal: 4%

This distribution reflects the ongoing global transition toward low-emissions electricity sources, with solar and wind continuing to grow rapidly due to falling costs and supportive policies, while fossil fuels remain dominant in many regions. The share of low-emissions sources, including renewables and nuclear, reached 41% of global electricity generation, a record high, and is expected to continue rising as countries pursue decarbonization targets and energy security goals [1].

There is a decrease in the share of fossil energies compared to 2011 (66.3%), in favor of solar and wind energies, which have more than doubled their share (4.6% in 2011). The share of nuclear power has slightly decreased (10.8% in 2011), while that of hydroelectric has remained stable (16.6% in 2011). The share of geothermal and bioenergy has increased by one point (2.5% in 2011).

Fossil Energy Production

Fossil fuel energies are derived from the transformation of organic materials that have been buried underground for millions of years. These include coal, oil, and non-renewable natural gas, which are rich in carbon-based hydrocarbons. In thermal power plants, these fuels are burned to produce electricity. The combustion process releases heat, which is used to convert water into steam. This steam then drives turbines connected to generators, producing electricity. Fossil fuels are the most widely consumed energy sources globally due to their high energy content and established infrastructure. However, they have significant drawbacks. They are non-renewable resources, meaning their reserves are finite and will eventually be depleted. Additionally, burning fossil fuels releases previously trapped carbon mainly in the form of carbon dioxide, which contributes to atmospheric pollution and exacerbates the effects of climate change.

Nuclear Energy Production

Nuclear energy is harnessed from the strong forces that bind protons and neutrons within the nucleus of an atom. This energy can be released through nuclear fission or fusion. In nuclear fission, the nucleus of a heavy atom, such as uranium, is split into two lighter nuclei, releasing a substantial amount of energy and additional neutrons. This process is utilized in nuclear power plants, where the heat generated from fission is used to produce steam, which then drives turbines to generate electricity. Nuclear energy is a potent source of power that does not emit greenhouse gasses during operation, thus reducing reliance on fossil fuels. However, it presents challenges such as safety concerns, the management of radioactive waste, and the risk of proliferation of byproducts that can be weaponized.

Renewable Energy Production

Renewable energies are derived from natural sources that replenish at a rate faster than their consumption. These sources include solar radiation, wind force, Earth's geothermal heat, water movement, and organic materials. Various technologies harness these energies to generate electricity. Photovoltaic (PV) solar panels convert sunlight directly into electricity, while wind turbines use the kinetic energy of wind to produce power. Hydroelectric plants exploit the potential energy of falling or flowing water to drive turbines, and geothermal plants utilize the Earth's internal heat to produce steam for electricity generation. Biomass energy is derived from organic materials, such as plant and animal waste, which can be burned directly for heat or converted into biofuels like ethanol and biodiesel. Renewable Natural Gas (RNG), also known as biomethane, is produced from the decomposition of organic waste in anaerobic digesters or landfills, capturing methane and purifying it for use as a substitute for conventional natural gas. Renewable energies are considered clean because they produce little to no net greenhouse gasses during operation, contributing to a diversified and secure energy supply. However, they face challenges such as intermittency (solar and wind are variable, while hydroelectric is generally stable and geothermal is constant), cost, and environmental impacts related to their integration into existing systems. Both biomass and RNG contribute to the renewable energy mix by providing versatile and dispatchable energy sources that can complement intermittent renewables like solar and wind. The sustainability of biomass and RNG depends on responsible sourcing and management practices to avoid deforestation, creating food shortages and other deleterious environmental impacts.

New Renewable Energies (NRE)

In recent years, the term "New Renewable Energies" (NRE), abbreviated as such, has been employed to differentiate renewable energy sources from more traditional and non-renewable sources such as coal, oil, and natural gas. The use of the term "new" is often associated with the transformation towards a more modern and sustainable approach to energy production, emphasizing newer technologies and practices within the framework of sustainable development and environmental protection.

The term "new" does not imply that these renewable energies have recently been discovered; rather, it signifies their perception as an emerging and progressive alternative to traditional energy sources. The adjective "new" thus considers these renewable sources that were either not exploited or minimally utilized until the 21st century.

According to the International Renewable Energy Agency (IRENA), the global distribution of RE shows that hydropower remains the primary renewable electricity-producing sector. Significant growth in renewable energies is concentrated in certain countries and regions such as Asia, the US, and Europe. It is noteworthy that nearly 59% of newly installed renewable capacities in 2022 were particularly deployed in Asia.

To illustrate recent shifts in power generation, we turn to data from Ember's *Global Electricity Review 2024* [2]. Global electricity generation has undergone a major transformation over the past decade, marked by a strong shift toward renewable energy and a gradual decline in fossil fuel use. In 2023, renewables accounted for 30% of global electricity generation, up from 19% in 2000. At the same time, the fossil fuel share fell from its peak of 68% in 2007 to 61% in 2023. Within fossil fuels, coal remained the largest source, generating 10,434 TWh or 35% of global electricity. Natural gas followed with 6,634 TWh (23%), while other fossil fuels contributed 786 TWh (2.7%).

Wind and solar power have been the fastest-growing sources of electricity, expanding from just 0.2% of the global mix in 2000 to 13.4% in 2023, producing 3,935 TWh. Solar reached 1,000 TWh of annual generation in just eight years, while wind took twelve—much faster than the growth trajectories of coal (32 years), gas (28 years), or hydropower (39 years).

Traditional low-carbon sources have seen a relative decline in share. Hydropower, though still the largest clean electricity source, dropped from 17% of the global mix in 2000 to 14% in 2023. Nuclear power fell from 16.6% to 9.1% over the same period. These changes reflect the pace of wind and solar deployment more than a decline in absolute output. Together, these trends highlight how technological progress, falling costs, and strong policy frameworks are driving the shift to clean electricity—what some analysts now call a new era of falling fossil fuel generation.

Solar Cost Trends and Microgrid Integration

The rapid emergence of NREs in the electricity production portfolio is revolutionizing the methods of electricity management and distribution at local, national, and international levels. Consequently, extensive deliberations are underway in numerous countries to establish a legal framework and harmonious integration of these new production means.

Cost reductions in solar photovoltaics have been central to this transition. Between 2010 and 2019, the global weighted-average levelized cost of utility-scale solar PV fell by 82%, dropping from USD 0.378 to USD 0.068 per kilowatt-hour. This was driven by falling module prices (down about 90%), declining balance-of-system costs, and improvements in efficiency and scale. Looking ahead, McKinsey projects that low-carbon sources—mainly solar and wind—could supply between 65% and 80% of global electricity by 2050, up from around 32% today.

The Solar Market Insight Report published by the Solar Energy Industries Association (SEIA) and Wood Mackenzie [3] provides detailed insights into the quarterly prices of solar photovoltaic (PV) multi-modules in the US from the first quarter of 2016 to the third quarter of 2020. According to the report, there is a clear downward trend in prices over this period, starting at approximately $0.70 per watt in Q1 2016 and decreasing to around $0.19 per watt by Q3 2020. This significant reduction in cost can be attributed to advancements in technology, increased production efficiency, and economies of scale in the solar industry. The consistent decline in prices highlights the growing affordability and accessibility of solar energy, which is crucial for the widespread adoption of renewable energy sources.

The Global Market Outlook for Solar Power 2023–2027 published by SolarPower Europe and the Global Solar Council [4] provides detailed forecasts for solar PV capacity in Italy. According to the report, Italy is expected to add 47 GW of new solar PV capacity by 2027. The medium scenario forecasts an addition of 28 gigawatts, while the low scenario predicts 23 GW. These projections highlight the potential growth in solar PV capacity in Italy.

This section draws on data from reports by Ember, Enerdata, McKinsey, IRENA, Solar Business Hub, and Mercom India to analyze global trends in renewable energy adoption, fossil fuel decline, and shifts in solar manufacturing [5]. In 2023, renewables generated 30% of global electricity—a ten-point increase since 2010 and the highest on record. Wind and solar drove this growth, rising from 0.2% of the global electricity mix in 2000 to 13.4% (3,935 TWh) in 2023, with solar delivering its fastest growth yet at 23% and wind at 10%. Despite this shift, fossil fuels still provided around 60% of worldwide electricity in 2023, mainly coal (35%) and natural gas (22%); other fossil fuels made up about 3%. Nuclear energy accounted for roughly 9%, and hydropower around 15%, though hydro output slipped due to widespread drought. Including renewables and nuclear, low-carbon sources reached nearly 40% of global electricity—an all-time high, signaling the start of what analysts call a turning point in the transition away from fossil fuels.

China's dominance in solar manufacturing also reshaped the global landscape. Solar module production costs in China dropped 42% in 2023, to USD 0.15 per watt—about 60% cheaper than U S prices—thanks to economies of scale and improved industrial capacity. Industry forecasts suggest that China will add approximately 300 GW/year of module production capacity over the next two years—comprising roughly 70% of 2023 global demand—driving further cost reductions and accelerating solar deployment worldwide.

From On-grid to Off-grid: A Journey towards Autonomy

Off-grid Systems

Off-grid systems are energy systems that operate autonomously, without being connected to a central electrical grid (or public electrical network). These systems are often used in rural, mountainous, or isolated areas where access to the electrical grid can be challenging or too costly.

Off-grid systems generally have three primary functions: energy production; conversion and utilization of this energy; and energy storage.

With off-grid microgrid systems, energy is typically produced from renewable sources, such as the sun for PV solar systems. Other energy sources can also be used, such as wind for wind turbines or the heat of the earth for geothermal systems or micro hydroelectric turbines in the 5–500 KW range. This energy will then be converted by a power electronics system (energy converter) to be either used directly in the system or stored in batteries (electric, mechanical, or thermal).

Here, we will mainly consider the use of PV solar and battery storage to illustrate our points in the context of microgrids.

We turn to data from Wood Mackenzie and the Center for Climate and Energy Solutions for a data analysis tracking U S microgrid growth [6].

Microgrid deployment in the United States has expanded rapidly in both scale and technological maturity. By the end of 2023, operational capacity reached 8.6 gigawatts, growing at an average annual rate of 32%, signaling a major shift in distributed energy infrastructure. As of early 2023, the country had 692 operational microgrids with a total capacity of approximately 4.4 gigawatts, concentrated primarily in Alaska, California, Georgia, Maryland, New York, Oklahoma, and Texas. This capacity represents about 82% of New York City's peak summer electricity demand of 10.5 gigawatts, illustrating the growing role of microgrids in supporting localized energy resilience.

This surge in deployment is fueled by evolving state and federal policy support, including grants and incentives, as well as the rise of microgrid-as-a-service business models that reduce barriers to entry for customers. The U S Department of Energy projects that total capacity could reach 20 GW by 2028. Once used mainly by military bases and remote areas, microgrids are now essential components of a modernized, resilient grid—especially as they integrate solar photovoltaics and battery storage, helping to ensure backup power during increasingly frequent climate-driven disruptions.

How Does Household Concrete Consumption Affect Microgrids?

In a quiet suburban neighborhood in Texas, the Smith family commences their day with a flurry of activity. The refrigerator operates continuously, maintaining the freshness of groceries and consuming approximately 235 kWh annually [7]. The kitchen is illuminated by a mix of energy-efficient LED bulbs, compact fluorescent lamps (CFLs), and incandescent bulbs, contributing to the household's annual lighting consumption of approximately 1,105 kWh [8]. As the day progresses, laptops and computers remain in use, with each device consuming approximately 20.24 kWh per year [9]. The dishwasher completes its daily cycle, using 1.2 kWh per load, while the washing machine, used three times a week, adds another 140.4 kWh to the yearly tally [10].

Meanwhile, across the Atlantic in a cozy apartment in Berlin, the Müller family follows a slightly different routine. The refrigerator, which is comparable in energy consumption to the Smiths', also uses approximately 235 kWh annually [11]. However, the Müllers are more conservative with their lighting, using approximately 800 kWh per year [12]. Computers and laptops are used infrequently, contributing to a slightly lower annual consumption of approximately 18 kWh per device [13]. The dishwasher and washing machine are used less frequently, reflecting a more prudent approach to energy consumption. The dishwasher uses approximately 1 kWh per cycle, while the washing machine consumes about 120 kWh annually [14].

These discrepancies in household energy consumption mirror broader tendencies in consumer behavior between Europe and the US. In the US, the prevalence of larger homes, reliance on central air conditioning instead of more efficient ductless systems used predominantly in Europe where air conditioning is needed and a higher number of electrical appliances contributes to a greater overall energy consumption rate. By contrast, European households are generally more energy-conscious, driven by higher energy prices and stringent efficiency standards.

The Smiths' energy consumption patterns are influenced by their need for heating and cooling. In Texas, air conditioning represents a significant proportion of overall load, particularly during the hot summers. The HVAC system is known to use up to 3,500 kWh annually [15]. Additionally, entertainment devices, including multiple TVs and gaming consoles, contribute an additional 1,200 kWh to their yearly consumption [16]. The Müllers rely on heating to a significant extent during the cold winters in Berlin. They utilize a combination of natural gas and electricity, with heating accounting for approximately 63.5% of their total energy consumption [17]. Germany's robust commitment to renewable energy sources has enabled the Smiths to benefit from solar panels and other green technologies, reducing their reliance on traditional energy sources [18].

American households, like the Smiths', tend to have larger homes with more appliances, leading to higher overall energy consumption. The average American household consumes about 10,972 kWh annually [19]. European households, such as the Müllers', are generally more energy-efficient due to higher energy costs and stricter regulations, with the average European household consuming around 3,500 kWh annually [20]. Americans typically use more energy-intensive appliances and have a higher number of devices per household. Europeans tend to use fewer appliances and prioritize energy efficiency. In the U S, air conditioning is a significant part of overall consumption, while in Europe, heating is. European homes also often use more efficient heating systems and insulation.

Energy policies in the U S vary by state, with a focus on both renewable and non-renewable energy sources. There is a growing trend towards energy efficiency, but the progress is uneven. The EU has stringent energy efficiency standards and a strong emphasis on reducing carbon emissions. Policies like the Energy Performance of Buildings Directive (EPBD) drive improvements in household energy efficiency [21].

Understanding these consumption patterns is crucial for designing effective microgrid systems. By tailoring energy solutions to the specific needs and behaviors of different regions, we can create more efficient and resilient energy infrastructures that cater to the unique lifestyles of families like the Smiths and the Müllers. Finally, the footnote lists numerous sources that provide average electricity consumption figures for common household appliances [22].

Development of the Off-grid Field

Solar Home Systems

The first off-grid systems arrived in the late 1980s, powering simple homes isolated from the electrical grid (such as mountain huts) to provide light this was known as the "solar-home-system".

This was the infancy of storing energy produced by solar panels in batteries, typically lead-based. This type of application quickly proved its worth, bringing energy where there was none and greatly improving living conditions for a small portion of the approximately 1.5 billion human beings without access to electricity. However, the market remained a niche one and solar technology did not spread widely, resulting in installations remaining very costly until the early 21st century.

Regarding the price of solar panels and the evolution of this technology, we can take the example of the first solar installation connected to the electrical grid in Europe (we leave off-grid for a moment for this illustration). The first European photovoltaic installation connected to the grid was inaugurated in 1982 in Lugano, Switzerland. This installation [23] is still in service and comprises 288 panels totaling a power of 10 kWp. By analyzing the cost of this installation, we can then estimate that the cost per kilowatt (kW) was about 21,000 CHF in 1982.

This section draws on data and analysis from NREL and IRENA to examine how the cost structure of utility-scale solar PV systems in the United States has evolved since 2017 [24].

Since 2017, the cost structure of utility-scale PV systems in the United States has changed significantly. According to NREL's Fall 2024 Solar Industry Update and Annual Technology Baseline, capital expenditure for a 100 MW utility-scale solar PV plant reached approximately USD 1.56 per watt AC in 2023. Module prices fell to around USD 0.10 per watt DC in the third quarter of 2024, close to historical lows, despite a slight 1% quarterly increase. These low module prices continue to carry a premium over global spot prices due to domestic manufacturing incentives and trade policies. The 68% decline in module prices since 2017 has been the main driver of overall system cost reductions, alongside progress in inverter technology and improvements in installation and design efficiencies.

According to the International Renewable Energy Agency, the global weighted average levelized cost of electricity for utility-scale solar dropped by about 70% between 2010 and 2023, from USD 0.39 to under USD 0.12 per kilowatt-hour. Total installed costs also fell by 79% to approximately USD 995 per kilowatt by 2019. In parallel, soft costs—such as permitting, financing, and project development—are receiving more attention as hardware costs decline. Ongoing efforts to simplify regulations and scale manufacturing are expected to further reduce total system costs, reinforcing solar PV as one of the most economically competitive technologies for clean electricity generation.

Hybrid System

With the development of SHS (Solar Home System) installations, as well as the evolution of power electronics components, technologies quickly emerged in the 1990s that combined the production of renewable solar energy (or other renewable sources such as wind or hydroelectricity) with fossil energy produced by generators.

Off-grid PV hybrid systems are installations that combine renewable energy, typically solar, and thermal energy. They can integrate a photovoltaic generator and another generator such as a wind turbine, a genset, a hydroelectric system, and sometimes even the public electricity grid (a practice that we will detail more broadly in the following topic—microgrids).

These systems offer numerous advantages, notably the optimization of solar energy. Unlike traditional storage systems with batteries, the hybrid inverter only stores energy when necessary. This allows for the avoidance of significant losses and the optimization of the use of solar electricity. Also, hybrid inverters associated with batteries provide a reliable solution to keep your domestic electrical network operating during a power outage. This advantage will become one of the major arguments for self-sufficiency many years later, with the advent of EVs. Hybrid installations will also introduce the concept of analysis and control of incoming and outgoing energies of a system and a more in-depth mastery of a kWh produced, stored, and used.

Let's look now at more data about microgrids. This analysis shows how microgrids are being used in different US states to improve energy resilience and sustainability.

According to Greentech Media, the penetration capacity of microgrids in various states across the US in 2020 shows significant variation. Texas leads with a microgrid capacity of 487 MW, followed closely by New York with 444 MW and California with 293 MW. Hawaii and Alaska also show substantial capacities, with 179 MW and 163 MW, respectively. Other states like Georgia, Michigan, Maryland, Arizona, and Oklahoma have capacities ranging from 160 to 81 MW [25].

This distribution highlights the varying levels of microgrid adoption across the country, influenced by factors such as state policies, energy needs, and the availability of renewable energy resources. Microgrids are crucial for enhancing energy resilience, especially in states prone to natural disasters or those with remote areas. They provide localized energy solutions that can operate independently from the main grid, ensuring a reliable power supply during outages. The data underscores the growing importance of microgrids in the U S energy landscape, driven by the need for sustainable and resilient energy systems.

According to GlobalData [26], the microgrid market share by region from 2018 to 2027 shows significant changes. In 2018, the Americas led the microgrid market with 54% of the share, followed by Asia-Pacific at 34% and EMEA at 22%. By 2022, the market shares shifted, with the Americas holding 43%, Asia-Pacific increasing to 38%, and EMEA rising to 28%. The forecast for 2027 indicates further changes, with the Americas at 34%, Asia-Pacific at 43%, and EMEA at 23%.

This data highlights the growing adoption of microgrids across various regions, with Asia-Pacific expected to experience the most significant growth. The increasing use of microgrids is driven by the need for reliable, resilient, and sustainable energy solutions, particularly in regions prone to natural disasters or with remote areas. Microgrids provide localized energy solutions that can operate independently from the main grid, ensuring a reliable power supply during outages. The data underscores the importance of microgrids in the global energy landscape, driven by technological advancements and policy measures aimed at enhancing energy resilience.

Equipped with this experience and new, commercially available technologies, it was only a matter of time before those who designed and implemented offgrid systems began to apply their best practices to aid traditional, on-grid electrical network systems.

This inflection point required the confluence of a significant number of different factors, including:

- The price of solar panels and batteries falling to the point where they are more a generally affordable commodity than a luxury.
- Availability of a storage technology that allows a certain number of cycles (charge/discharge) suitable on a time scale of use.
- Energy markets are evolving. Fossil fuel prices are influenced by factors such as extraction costs, technological advancements, and regulatory environments. In the United States, the average production cost of natural gas has been below $3 per million British thermal units (MMBtu) in recent years. However, market prices have fluctuated, with the Henry Hub spot price averaging $4.19/

MMBtu in February 2025[27]. Crude oil production costs vary significantly by region and extraction method. In the Middle East, onshore production remains the most cost-effective, with breakeven prices averaging around $17 per barrel. In contrast, US shale oil projects have higher breakeven prices, averaging approximately $45 per barrel [28]. As of April 3, 2025, the price of West Texas Intermediate (WTI) crude oil is approximately $66.76 per barrel. These figures illustrate that the fossil fuel industry faces ongoing economic pressures due to factors such as the depletion of easily accessible reserves, the need for advanced extraction technologies, and evolving regulatory and environmental considerations.

- The promulgation of energy policies aimed at reducing greenhouse gas emissions encouraging residential and industrial electrification and energy resilience.

The adoption of electric transport is the most visible and impactful aspect of this transformation and sets a powerful precedent for the wider electrification of residential and industrial consumption, notably for heating, cooling, and cooking. This transformation is very positive because the energy to power most consumers, including their transport, can be produced affordably via solar power, which is carbon-neutral and can easily be paired with storage solutions.

It is important to note that the IEA predicts global energy consumption will rise by 1% per year until 2040, with a growth in electricity of 70% between 2013–2040 [29]. In the next few years, we can also expect a larger increase in the amount of electricity needed due to the quickly rising use of AI. AI data centers require much more power to operate than traditional data centers and are being very quickly built all over the world.

Microgrid Systems

Microgrids are a combination of the hybrid systems previously presented, but on a larger scale and with a certain organization and distribution of the energies produced (see example of diagram below). They are particularly effective in areas with high population density and high per capita electricity demand. The efficient use of energy is integral to the proper functioning of a system, enabling it to operate autonomously for 60–85% of the time if required, and plays a crucial role in improving access to electricity, especially in rural and isolated regions.

The basic principles of such networks rely on energy sharing and partial pooling of produced, stored, and consumed energies to support the central unit that will provide the majority of the system's energy. Microgrids are a perfect example of the functionalities that a traditional electrical network should offer today if it were a smart grid.

Figure 2.1 shows how a smart home energy management system works. It includes a smart home with devices, appliances, solar panels, and traditional grid power. The system also has battery storage. The Energy Management System

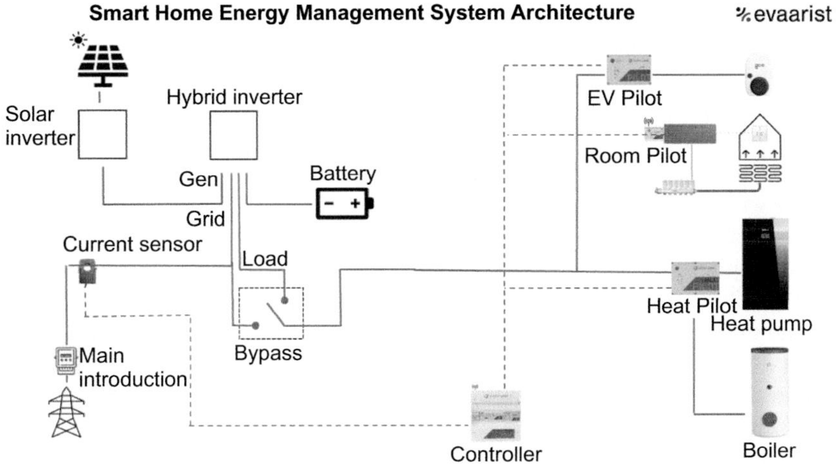

Fig. 2.1 Smart Home Energy Management System Architecture [30].

(EMS) optimizes energy usage, storage, and trading. The EMS manages energy in the home and trades excess energy with the grid or other homes to save money.

To underscore the importance of microgrids, it can be shown that they have enabled access to electricity in areas that traditional methods could not effectively or economically reach.

According to World Bank data, in 2021, 91.4% of the world's population had access to electricity. This figure has significantly evolved since the late 1990s, thanks notably to photovoltaic energy and the principle of microgrids. Over the last decade, microgrids have significantly contributed to increasing global electricity access, particularly in remote and underserved areas. According to the World Bank, the number of people connected to mini grids, which include microgrids, more than doubled from 5 million in 2010 to 11 million in 2019 [31]. This growth is driven by the declining costs of key components, such as solar panels and batteries, and the introduction of new digital solutions that enhance the efficiency and reliability of microgrids. Moreover, solar microgrids could provide uninterrupted, high-quality electricity supply to nearly half a billion people living in areas previously deprived of electricity and constitute a cost-effective solution to bridge the energy access gap by 2030. Microgrids provide communities off the main grid with high-quality, uninterrupted electricity at a low cost. The cost of solar mini-grids has dropped from $0.55 per kWh in 2018 to $0.38 per kWh today. This makes microgrids an attractive option for expanding electricity access. Their continued deployment is essential for achieving universal electricity access by 2030 [32].

We turn to a data analysis related to electricity access based on recent findings by the World Bank and Sustainable Energy for All [33].

From 1990 to 2022, global electricity access improved at an unprecedented pace, marking a major infrastructure achievement. The number of people without electricity fell from 1.14 billion in 2010 to 685 million in 2022, driven largely by gains in Asia, particularly in India and Bangladesh. However, for the first time in over a decade, this progress reversed: 2022 saw an increase of 10 million people without electricity access compared to 2021. This reversal was primarily due to population growth outpacing new connections, especially in Sub-Saharan Africa, where 570 million people—more than 80% of the global total without access—remain without reliable electricity.

Sustainable Energy for All reports that Africa has seen the smallest absolute gains, despite improving its electricity access rate from 44% to 59% between 2010 and 2022. Nigeria alone accounts for 86 million people without electricity, the highest national total globally. Today, 80% of people lacking electricity live in just 24 countries—20 of which are in Africa. If current policies continue, 660 million people are projected to remain without electricity by 2030, 85% of them in Sub-Saharan Africa.

To meet the UN Sustainable Development Goal of universal electricity access by 2030, the international community must urgently scale up decentralized energy solutions such as microgrids, mini-grids, and standalone solar systems. These are essential for reaching remote and rural populations—where eight in ten people without electricity currently live—and for addressing broader issues tied to energy poverty, including healthcare access, education, and gender inequality.

In Figures 2.2 and 2.3, we can see the evolution on the world map of the access to electricity in percentages for each country.

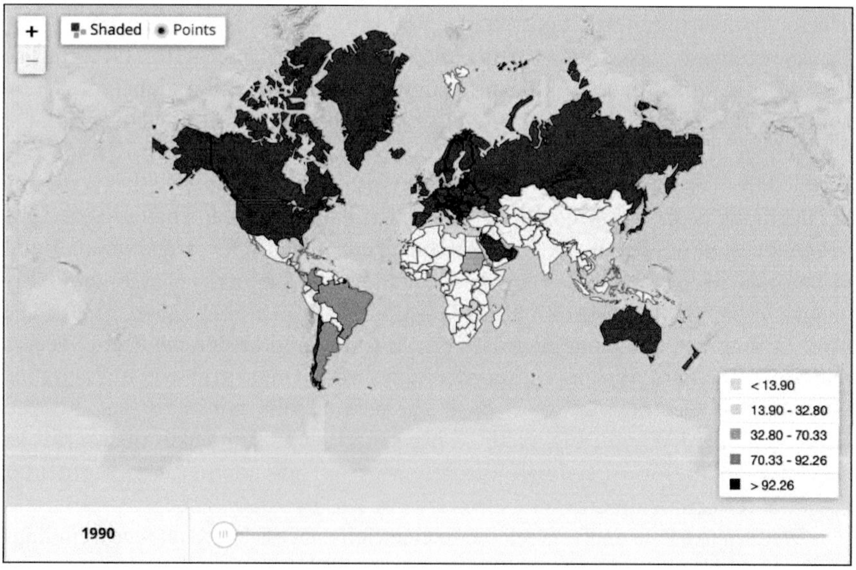

Fig. 2.2 World Map with Percentages of Access to Electricity in 1990 [34].

Fig. 2.3 World Map of Countries with Access to Electricity in 2022 (in Percentages) [35].

Autonomy: Operating Principle

The operating principle of a self-sufficient system is relatively simple and is largely inspired by the techniques previously presented for offgrid systems. A smart, self-sufficient solar power installation with battery storage allows one to conserve unused energy and gain the related flexibility of use. The system would switch automatically to "island" mode in case of power outage. An autonomy rate of 85% is easily achievable. This would mean that 85% of your annual electricity supply is provided by the sun!

Difference between Self-consumption and Autonomy

A microgrid system with a hybrid energy storage system can help avoid blackouts. The microgrid can work on its own or with the main grid. It uses solar panels and wind turbines, as well as regular power sources. The hybrid energy storage system, which includes batteries and supercapacitors, makes the microgrid more resilient and efficient. The Model Predictive Control (MPC) strategy makes sure the microgrid is managed and operated efficiently and can handle disruptions. The MPC coordinates energy sources and storage to keep the microgrid stable and efficient. It prevents blackouts and ensures reliable power even during bad weather. This resilience is important for keeping the power on in critical situations, which helps make the energy system more reliable and sustainable.

To understand how these systems work in everyday life, it's useful to explain the most important parts and how they're measured. Here are some short definitions of the main terms used when talking about solar-powered household systems:

- **Solar Panels:** These panels generate electricity from sunlight, providing a renewable energy source for the home.
- **Batteries:** Excess energy produced by the solar panels is stored in batteries for later use, ensuring that energy is available even when solar production is low.
- **Household Consumption:** This refers to the energy used by various appliances and devices within the home, such as lighting, heating, cooling, and electronics.
- **Grid Connection:** The home is connected to the main power grid, allowing it to buy additional energy when solar production is insufficient (e.g., at night or during cloudy days). Conversely, when the solar panels produce more energy than the household needs, the excess energy can be sold back to the grid.
- **Self-Consumption Rate:** This is the proportion of solar energy produced that is directly used by the household. It is calculated by dividing the energy used from solar by the total solar energy produced.
- **Autonomy Rate:** This indicates how much of the household's total energy consumption is supplied by its own solar panels. It is calculated by dividing the energy used from solar production by the total energy used by the household.
- **Energy Resold:** This refers to the excess energy produced by the solar panels that is sold back to the grid.
- **Energy Self-Consumption:** This is the portion of energy produced by the solar panels that is consumed within the household, rather than being stored or sold.

The self-consumption rate is the amount of energy produced by solar power that is used by the household. It can be expressed as:

Self-Consumption Rate = Energy Consumed from Solar Production/ Total Solar Energy Produced

This shows how much of the household's energy is produced by itself. It can be expressed as:

Autonomy Rate = Energy Consumed from Solar Production/ Total Energy Consumption Energy Resold

Energy resold is excess energy produced by the solar power system that is sold back to the grid. It can be expressed as:

Energy Resold = Total Solar Energy Produced - Energy Consumed from Solar Production

It is the same as the energy consumed from solar production. It can be expressed as:

Energy Self-Consumption = Energy Consumed from Solar Production

These indicators help assess a household's energy system, showing the benefits of using renewable energy sources like solar power. The autonomy rate shows how much energy a dwelling produces and uses. This shows how much energy can be produced by a building's own energy sources compared to how much it uses.

In a smart home, various energy sources and appliances interact to optimize energy usage and efficiency. The primary components involved are solar panels, batteries, household consumption, and the grid connection (Figure 2.4).

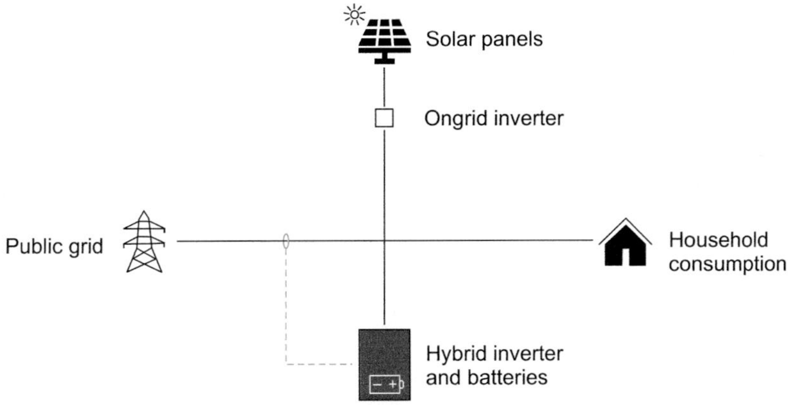

Fig. 2.4 Smart home connected to the grid [36].

These indicators help assess the efficiency and independence of a household's energy system. The autonomy rate, in particular, shows the potential for self-sufficiency by indicating how much energy a dwelling produces and uses from its own solar panels.

Experience Feedback

Following over two years of experience and hundreds of installations commissioned, the feedback is unambiguous. The implementation of a comprehensive energy strategy, which includes total roof coverage with solar panels, the installation of batteries and the electrification of consumption, particularly in mobility, has proven highly effective.

In the majority of cases, a policy of total roof coverage through solar panels allows households to harness the maximum amount of solar energy. Furthermore, the installation of batteries allows excess energy generated during the day to be stored for use at night or on cloudy days. This configuration not only guarantees a continuous power supply but also enhances energy autonomy. The integration of these systems enables households to achieve autonomy rates of 65–80%, including mobility. This results in the majority of energy consumed by these households being self-produced, which significantly reduces their reliance on the public electricity grid. They are also protected against increases in public network electricity tariffs, and do not depend on value-of-power fluctuations either.

The electrification of consumption, particularly in the context of mobility, serves to further amplify these benefits. The use of EVs charged with self-generated solar power contributes to overall energy independence and reduces carbon footprints. It has been demonstrated that households with integrated solar and battery systems, in conjunction with EVs, achieve significant savings on energy costs. These systems protect households from fluctuations in public network electricity tariffs and the volatility of power prices. Furthermore, they have a notable environmental impact. By reducing dependence on fossil fuels and minimizing greenhouse gas emissions, these households contribute to broader climate goals. The transformation to renewable energy sources, supported by advancements in solar technology and battery efficiency, is a critical step towards sustainable living. Extensive data supporting these findings is available from reliable sources such as the IEA and the U S Department of Energy. As an example, the IEA's report on rooftop solar PV systems indicates that the adoption of such systems is expected to quadruple by 2030, reaching approximately 100 million households globally [37]. Furthermore, the U S Department of Energy has identified the growing adoption of microgrids and distributed energy resources, which are vital for improving energy resilience and sustainability [38].

In summary, the integration of solar panels, battery storage, and electrified consumption, particularly in mobility, provides a comprehensive solution for achieving high levels of energy autonomy. This approach offers economic benefits by protecting against tariff increases and supports environmental sustainability by reducing reliance on non-renewable energy sources.

At the level of the annual energy balance, thanks to the complete solar roof, most users have a positive production/consumption ratio. This means that over the course of a year, they generate more electricity than they consume. However, their energy autonomy is not strictly above 100%, as they may still draw from the grid at times when solar production is low, such as at night or during extended cloudy periods.

In Figure 2.5, we present the energy data for a day in October—almost 100% autonomous operation, with 32.43% of electricity supplied by solar production and 65.77% from the battery. This graph shows how a self-sufficient energy system behaves over a 24-hour period.

The State of Charge (SOC) of the battery: (1) steadily decreases during the night as stored energy is consumed. When the sun rises (2) and the primary load is deactivated around 9:00 A.M., the battery begins to recharge. The system's electricity consumption (3) is visible throughout the day, including the activation of the heat pump at 3:00 A.M., which is powered entirely by the battery.

In section (4), we see electricity being exported to the grid—a result of excess solar production once the battery is fully charged. In (5), solar power is being used to recharge the battery, illustrating the system's efficient use of available solar energy.

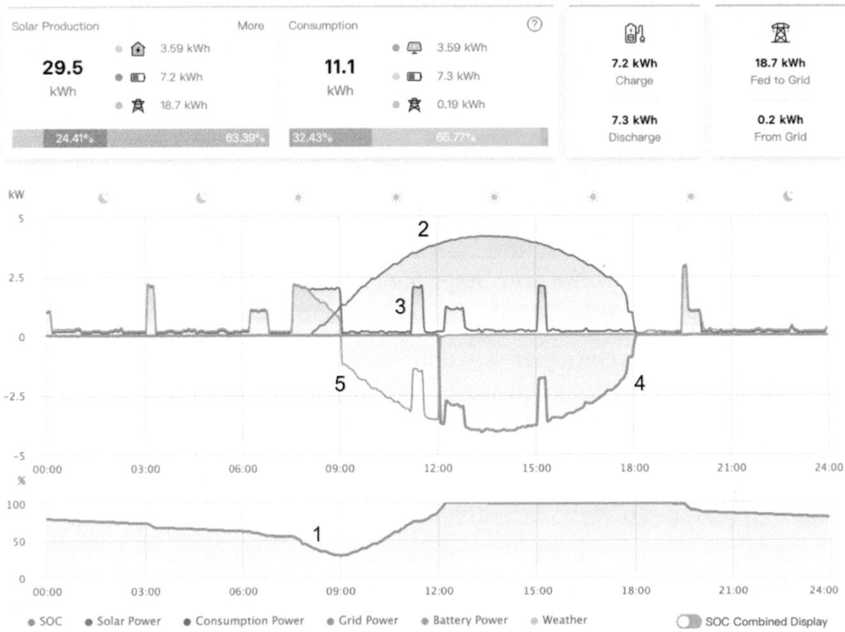

Fig. 2.5 Dashboard of a day of consumption [39].

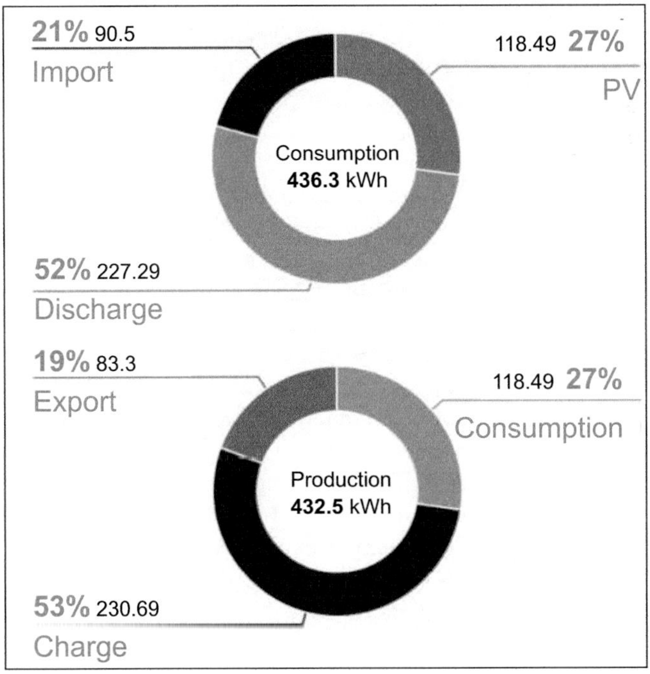

Fig. 2.6 Monthly energy balance of an autonomous installation: Production.

Monthly energy balance of an autonomous installation is seen in Figure 2.6 in two graphs: the first represents consumption, and the second shows production. Both include the notion of battery activity (charge/discharge), accounting for 52–53% of the energy. This represents the energy stored in the battery and subsequently used. Next is the PV/Consumption which represents 27% of the energy, referring to the energy produced by the solar panels and consumed directly. Finally, there are interactions with the electrical grid.

Own Consumer Networks and Self-consumption Communities: Towards Decentralized Energy

The concepts of "own consumer networks" and "self-consumption communities" are innovative approaches to greater energy independence that can significantly contribute to the transformation towards decentralized and sustainable energy systems. These initiatives empower individuals and groups to produce, consume, and share their own renewable energy, thereby reducing their reliance on large, traditional energy networks.

An "own consumer network" consists of individuals or communities who generate their own energy from renewable sources, such as solar or wind power, and share it locally through a communal grid. By producing and consuming energy locally, these networks facilitate more efficient energy management, reducing losses associated with long-distance transmission. This not only enhances overall energy efficiency but also significantly cuts greenhouse gas emissions, thereby minimizing the carbon footprint of the participants. Furthermore, these networks provide greater energy independence, enabling members to better control their energy costs and shield themselves from price volatility in the broader energy market.

Self-consumption communities (Figure 2.7), take this concept a step further by bringing together multiple households or businesses to collectively share locally produced energy, often generated through communal solar or wind installations. This energy is consumed on-site, thereby limiting dependence on the grid and optimizing energy efficiency. By pooling resources and infrastructure, members of these communities can reduce installation and maintenance costs while maximizing the utilization of the energy produced. Moreover, these communities strengthen social bonds and promote equitable resource sharing, while also raising awareness of energy and environmental issues among participants.

Although these concepts are still relatively new and not widespread, they represent a potential paradigm shift in how energy is produced and consumed. They offer a pathway towards greater sustainability by fostering energy autonomy and enhancing the resilience of communities in the face of future energy challenges.

Demand-Side Management (DSM)

Demand Side Management (DSM) plays a pivotal role in optimizing energy use within a microgrid community [40]. By managing energy consumption on the

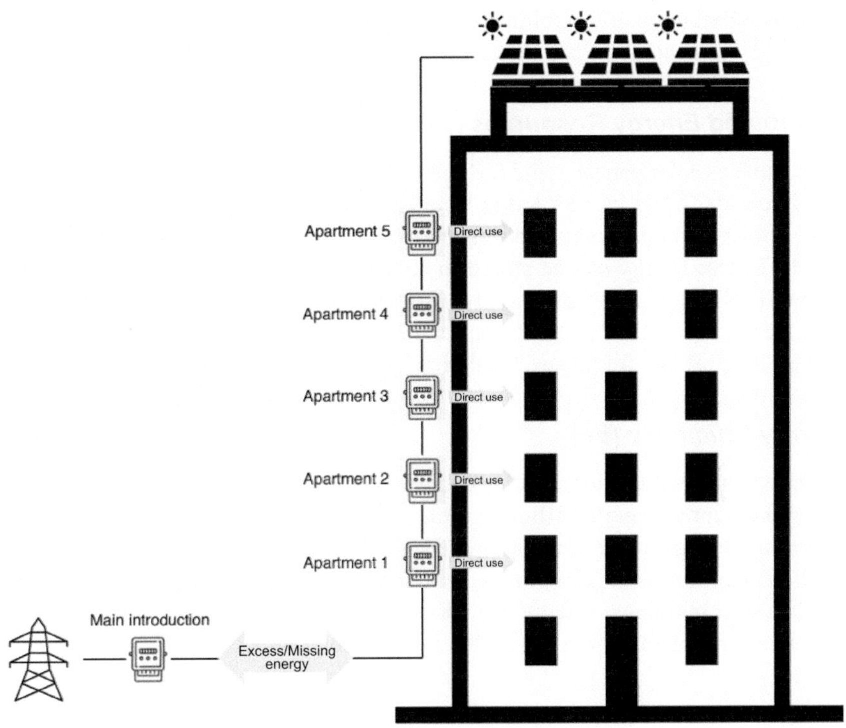

Fig. 2.7 Self-consumption network.
(*Source:* Viret Loïc, *Evaarist SA*.)

user side, DSM enhances the performance and efficiency of microgrids. This is achieved through continuous monitoring of energy use and strategic scheduling of devices, which helps reduce energy purchase costs and avoid penalties.

In the context of microgrids, DSM encompasses three main activities: demand response (DR), DER, and energy efficiency (EE). These activities collectively help microgrids manage demand more effectively, reducing peak loads and operational costs. By optimizing scheduling of energy use, DSM can defer the need for new power plants and leverage real-time pricing strategies to balance demand within smart grids.

Demand Response (DR)

DR involves adjusting the demand for power instead of adjusting the supply. In a microgrid, DR can be used to shift or reduce energy use during peak periods, thereby alleviating stress on the grid and reducing costs. For instance, during times of high demand, a microgrid can temporarily reduce the power consumption of non-essential devices or shift their operation to off-peak times. This not only helps in maintaining grid stability but also allows consumers to benefit from lower energy prices in some jurisdictions during off-peak periods.

Distributed Energy Resources (DER)

DER, such as solar panels, wind turbines, and battery storage systems are integral to microgrids. DSM in microgrids leverages these resources to manage demand more effectively. For example, during periods of high energy production from solar panels, excess energy can be stored in batteries and used later when production is low. This ensures a steady supply of power and reduces reliance on the main grid. Additionally, DERs can be coordinated to provide ancillary services such as voltage support and frequency regulation, further enhancing grid stability.

Energy Efficiency (EE)

EE measures are crucial for reducing overall energy consumption within a microgrid. By implementing energy-efficient technologies and practices, microgrids can lower their energy demand and improve sustainability. This includes using energy-efficient appliances, optimizing heating and cooling systems, and improving insulation in buildings. Energy efficiency not only reduces the operational costs of a microgrid but also minimizes its environmental impact.

Integration and Benefits

The integration of DSM with microgrids leads to more efficient and sustainable energy systems. This synergy allows for better utilization of renewable energy sources, improved energy efficiency, and potentially lower costs for consumers. By managing demand effectively, microgrids can operate more autonomously, reducing their dependence on the main grid and enhancing energy resilience. This is particularly beneficial in areas prone to power outages or in remote locations where grid access is limited.

Moreover, the ability to generate and manage energy locally empowers communities and businesses to take control of their energy needs [41]. This not only fosters a sense of energy independence but also promotes local economic development through job creation in the renewable energy sector. In the long term, the adoption of microgrids with integrated DSM can contribute significantly to the global effort to mitigate the impacts of climate change by reducing greenhouse gas emissions.

Return on Investment

If we consider a system as a whole and integrate the full energy costs of a single-family home, including the energy costs related to private mobility (excluding here public transport or air travel), it becomes clear that electrifying its consumptions (heating, cooling, lighting, cooking, cleaning, and mobility) while increasing its capacity for production and storage of self-produced electrical energy improves a project's return on investment, even though the investment required is larger.

In most cases, 60% of the savings are realized on the costs of purchasing electricity from the network (including a resale of a certain amount) and 40% are realized on the costs related to the purchase of fuel for private mobility—a purchase that disappears when using self-produced electricity for daily travel in a vehicle. Not included here are the purchase costs related to the electric vehicle since these are offset by the purchase costs of a thermal vehicle, because the variance between those prices is currently in flux but gradually approaching parity.

The example presented below, in Switzerland, also shows the important role played by public subsidies and tax gains achieved in the case of investment in renewable energy (whether the object is renovated or a new construction). Note that the direct subsidy part represents about 15% of the gross investment in this country. Tax savings are about 20%. This is a strong incentive mechanism. Also note the case below presents a saving made on replacing a roof or in new construction since the solar panels replace the classic tiles. And that represents quite an amount!

How to Further Improve the Return on Investment Time

Electricity pricing methods are changing. In Europe and around the world, new pricing methods are being developed, including dynamic pricing. Dynamic pricing means electricity prices change based on supply and demand. This approach aims to show the real cost of making electricity and encourage people to use less.

In Europe, dynamic pricing is becoming more common because of new smart grid technology and government support. The European Union's Clean Energy Package says that all member states must offer consumers dynamic electricity price contracts if they ask for them. The idea of a smart grid is to use digital and other advanced technologies to monitor and manage the transport of electricity from all sources to meet changing electricity demands. Smart grids allow utility providers and consumers to communicate with each other. This allows for the exchange of real-time data and more efficient energy management. Smart grids use technology like smart meters, sensors, controls, and data centers to improve energy distribution, reliability, and the environment. By using these technologies, smart grids can work better with everyone involved in the energy system, including generators, grid operators, and end-users. This helps make the energy system stronger and more sustainable. These contracts link retail prices to wholesale prices, which change throughout the day. In Europe, dynamic pricing includes several models [42]. One is Time-of-Use (ToU) Pricing, where prices vary at different times of the day. CPP charges more during periods of high demand. RTP changes prices in real time based on wholesale market conditions. These models encourage people to use less electricity at peak times, which helps the grid and lowers energy costs. Dynamic pricing is also becoming popular around the world. In the United States, the Federal Energy Regulatory Commission (FERC) is promoting dynamic pricing to improve grid efficiency and reduce peak demand. Smart meters and AMI help make dynamic pricing possible. They collect and share data in real time between utilities and consumers.

Dynamic pricing affects how long it takes to recoup the cost of traditional solar installations [43]. Solar panels make electricity during the day, when electricity prices are highest. This can help solar investments make more money by saving on electricity bills. The payback period of solar installations depends on electricity rates, government incentives, and battery storage. Higher electricity prices during peak periods save more money from solar power and pay for the system faster. Programs like the Smart Export Guarantee (SEG) in the UK can reduce the payback period. Battery storage lets households save more and pay back their investment faster.

In Europe [44], solar energy is being adopted faster thanks to dynamic pricing and supportive policies. The EU wants to be carbon neutral by 2050. It has set intermediate targets, such as cutting CO_2 emissions by 55% by 2030. This ambitious plan is leading to more investments in renewable energy, including solar power. Dynamic pricing is changing the electricity market [45]. It makes energy use more efficient and makes renewable energy investments more profitable thereby helping Europe and the world move towards a sustainable energy future. Dynamic pricing also benefits consumers and supports climate goals by reducing reliance on fossil fuels and minimizing greenhouse gas emissions.

Meanwhile, in the United States, the passage of the Inflation Reduction Act (IRA) has made available a wide range of generous tax incentives and credits available to companies, local and state governmental entities and individual power consumers to encourage the development of clean, resilient power. Additional incentives are provided by many states, and the authors recommend that interested parties research what might be available to them in their respective states.

Below are examples of specific IRA incentives that were available to microgrid developers as of July 2025. Unfortunately, recent legislation in US Congress reduced or eliminated many of these. Please be sure to confirm the availability and details of any of these tax credits when planning your project, as there is a high probability these will be modified, scaled back, or repealed. The full list can be found in the source link provided for this information [46].

A Personal Guide to Transformation: Essential Golden Rules

As we navigate the complexities of the global energy transformation, it is crucial to recognize that individual actions play a pivotal role in shaping a sustainable future. By adopting certain principles in our daily lives, we can contribute meaningfully to reducing our environmental footprint and advancing the global shift towards renewable energy. This chapter outlines a personal guide to energy transformation, highlighting a few key practices that can help individuals make a significant impact.

Table 2.1 Selected IRA Incentives for Microgrid Developers (as of June 2025)

Credit Name	*Credit Description*	*Eligible Recipients*	*Base Credit Amount*	*Bonus Credit Amount*
Production Tax Credit for Electricity from Renewables	Provides a tax credit for production of electricity from renewable sources.	Facilities generating electricity from wind, biomass, geothermal, solar, small irrigation, landfill and trash, hydropower, and marine and hydrokinetic renewable energy.	0.3 cents/kW, inflation adjusted	Credit is increased by 5 times for projects meeting prevailing wage and registered apprenticeship requirements. Credit is increased by 10% if the project meets certain domestic content requirements for steel, iron, and manufactured products. Credit is increased by 10% if located in an energy community.
Investment Tax Credit for Energy Property	Provides a tax credit for investment in renewable energy projects.	Fuel cell, solar, geothermal, small wind, energy storage, biogas, microgrid controllers, and combined heat and power properties. For solar, includes: (1) equipment that uses solar energy to generate electricity, to heat or cool (or provide hot water for use in) a structure, or to provide solar process heat, and (2) equipment that uses solar energy to illuminate the inside of a structure using fiber-optic distributed sunlight or electrochromic glass that uses electricity to change its light transmittance properties in order to heat or cool a structure.	6% of qualified investment (basis of energy property)	Credit is increased by 5 times for projects meeting prevailing wage and registered apprenticeship requirements. Credit is increased by up to 10 percentage points for projects meeting certain domestic content requirements for steel, iron, and manufactured products. Credit is increased by up to 10 percentage points if located in an energy community.

Contd.

Table 2.1 *Contd.*

Credit Name	*Credit Description*	*Eligible Recipients*	*Base Credit Amount*	*Bonus Credit Amount*
Low-Income Communities Bonus Credit	Provides an additional investment tax credit for small-scale solar and wind facilities on Indian land and in low-income communities.	Solar and wind facilities with a maximum net output of less than 5 MW, including associated energy storage technology.	6% of qualified investment (basis of energy property)	Credit is increased by 10 percentage points for projects located in low-income communities or on Tribal land. Credit is increased by 20 percentage points for projects that are part of certain federally subsidized housing programs or that offer at least 50% of the financial benefits of the electricity produced to low-income households. This bonus amount will require an application by the taxpayer, with a cumulative total of 1.8 GW of direct current capacity per year available for allocation.
Clean Electricity Production Tax Credit	Provides a technology-neutral tax credit for production of clean electricity. Replaces the production tax credit for electricity generated from renewable sources (§45) for facilities placed in service in 2025 and later.	Facilities generating electricity for which the greenhouse gas emissions rate is not greater than zero.	0.3 cents/ kW [inflation adjusted]	Credit is increased by 5 times for projects meeting prevailing wage and apprenticeship requirements. Credit is increased by 10% for projects meeting certain domestic content requirements for steel, iron, and manufactured products. Credit is increased by 10% if located in an energy community.

Contd.

Table 2.1 *Contd.*

Credit Name	*Credit Description*	*Eligible Recipients*	*Base Credit Amount*	*Bonus Credit Amount*
Clean Electricity Investment Tax Credit	Provides a technology-neutral tax credit for investment in facilities that generate clean electricity. Replaces the investment tax credit for energy property (§48) for property placed in service in 2025 and later.	Facilities that generate electricity with a greenhouse gas emissions rate that is not greater than zero and qualified energy storage technologies.	6% of qualified investment (basis)	Credit is increased by 5 times for facilities meeting prevailing wage and apprenticeship requirements. Credit is increased by up to 10 percentage points for facilities meeting certain domestic content requirements for steel, iron, and manufactured products. Credit is increased by up to 10 percentage points if located in an energy community.
Energy Efficiency Home Improvement Credit	Provides a tax credit for energy-efficiency improvements of residential homes.	Homeowners; Renters for certain improvements	30% of cost, with limits for each type of improvement and total per year. Credit capped at $600 for "energy property," e.g., efficient heating and cooling equipment; $600 for windows; $250 per door, $500 total for doors; $2,000 for heat pumps; $1,200 for qualified energy efficiency improvements to the building envelope, including insulation and air sealing. Total annual credit capped at $1,200, with a separate annual $2,000 limit for heat pumps. $150 credit for home energy audits.	None

Contd.

Table 2.1 *Contd.*

Credit Name	*Credit Description*	*Eligible Recipients*	*Base Credit Amount*	*Bonus Credit Amount*
Residential Clean Energy Credit	Provides a tax credit for the purchase of residential clean energy equipment, including battery storage with capacity of at least 3 kWh.	Homeowners (including renters)	30% of cost of equipment through 2032; 26% in 2033; 22% in 2034.	None
New Energy Efficient Homes Credit	Provides a tax credit for construction of new energy efficient homes.	Homebuilders	$2,500 for new homes meeting Energy Star standards; $5,000 for certified zero-energy ready homes. For multifamily, base amounts are $500 per unit for Energy Star and $1000 per unit for zero-energy ready.	For multifamily homes, 5 times the base amount if prevailing wage requirements are met.
Energy Efficient Commercial Buildings Deduction	Provides a tax deduction for energy efficiency improvements to commercial buildings, such as improvements to interior lighting; heating, cooling, and ventilation, hot water; and building envelope.	Owners and long-term lessees of commercial buildings. Designers of energy efficient building property (architects, engineers). Tax-exempt owners of commercial properties, pending Treasury guidance on deduction allocation.	$0.50–$1 per square foot, depending on increase in efficiency, with deduction over four year periods capped at $1 per square foot. Inflation adjusted. Alternatively, taxpayers can deduct adjusted basis in "qualified retrofit plans" that reduce a building's energy use intensity by at least 25%.	5 times the base amount if the project meets prevailing wage and apprenticeship requirements.

1. Drive Electric: Embracing Sustainable Transportation

One of the most straightforward and impactful steps individuals can take is switching from fueled to EVs. Microgrids will in the future be designed to accommodate EVs, and pairing electrified transport with a microgrid also enhances energy autonomy by ending reliance on supply of fuels by the oil and gas industries and redirecting the value of those avoided costs back to the consumer. The transportation sector is also a major contributor to greenhouse gas emissions and poor air quality, and by choosing to drive electric, we can substantially reduce our carbon footprint and improve the quality of the air we breathe.

EVs are good for the environment because they do not produce any emissions when operated. This reduces CO_2 and other pollutants in the air, making it cleaner and reducing greenhouse gases [47]. As EVs become more affordable, they are becoming more competitive with traditional cars. Many countries and regions give tax rebates and lower registration fees to people who buy electric cars which makes them an attractive option for consumers [48]. Driving EVs helps us be more energy independent especially when charging your EV with home-produced solar power. This makes us less dependent on purchasing fossil fuels and therefore more secure at home [49].

2. Produce What You Consume: Self-sufficiency in Energy

A key principle of the energy transformation at the individual level is to produce and store as much of your own energy as possible. Whether it is through solar panels, wind turbines, or other renewable energy sources, generating your own power is a critical step toward sustainability.

- **Solar Power:** Installing solar panels on your roof is one of the most effective ways to produce your own electricity. Depending on your geographic location and the size of your installation, you may be able to generate enough power to meet all of your household's energy needs.
- **Wind Energy:** For those living in areas with consistent wind patterns, small-scale wind turbines can be an excellent complement to solar power, providing energy during periods when the sun isn't shining.
- **Battery Storage:** To maximize the benefits of self-produced energy, consider investing in a home battery storage system. This allows you to store excess energy produced during the day and use it at night or during periods of low generation.

3. Consume the Type of Energy You Can Produce

An often-overlooked aspect of energy self-sufficiency is aligning your energy consumption with the type of energy you can produce. This means prioritizing the use of electricity, which can be generated from renewable sources, over other forms of energy that are typically derived from fossil fuels.

- **Electric Appliances:** Shift towards using electric appliances for heating, cooking, and other household needs. For example, replacing a gas stove with an electric or induction model not only reduces reliance on fossil fuels but also allows you to directly utilize the renewable electricity you generate and receive the benefit of the avoided costs you would be paying to the fossil fuel supplier.
- **Electric Heating:** Consider switching to electric heating systems, such as heat pumps, which are highly efficient and can be powered by renewable energy.
- **Smart Consumption:** Utilize smart home systems to optimize energy use, ensuring that high-energy tasks are performed when your energy production is at its peak, further aligning consumption with production.

A smart home has many connected devices and systems. You can control them remotely, often through a central hub or smartphone app. These devices include smart thermostats, lighting, appliances, and security systems. They can work together to improve energy efficiency, convenience, and security. Use these systems to make the most of your energy use. This means that high-energy tasks will be done when you have the most energy, which best matches what you make with what you use.

4. Maximize Your Autonomy: Striving for Energy Independence

Achieving a high degree of energy autonomy is the ultimate goal for individuals looking to minimize their environmental impact and increase their resilience in the face of energy uncertainties.

- **Energy Efficiency:** Before expanding your energy production capabilities, focus on reducing your overall energy consumption. This includes investing in energy-efficient appliances, improving home insulation, and adopting energy-saving habits.
- **Grid Independence:** While staying connected to the grid offers security, particularly during periods of low renewable energy production, strive to reduce your reliance on external energy sources as much as possible. This not only reduces your carbon footprint in most jurisdictions but also shields you from potential price increases or energy shortages.
- **Community Initiatives:** Consider participating in or establishing local energy cooperatives or microgrids, where you and your neighbors can share energy resources, further enhancing collective energy independence.

The path to a sustainable future is paved with individual actions. By adopting these essential practices—electrifying transport and other energy consumption, producing what we consume, aligning our energy use with our production capabilities, and maximizing our energy autonomy—we can each play a crucial role in the global energy transformation. While these steps may require upfront investments and lifestyle changes, the long-term benefits for both the planet and our personal well-being are undeniable. As more individuals embrace these principles, we move closer to a world where clean, renewable energy is the norm rather than the exception.

Overview of the Possible Renewable Energy Sources and Storage Technologies with a 20-year Trend for Each Technology

In this section, we will provide information you need about different types of renewable energy and how they are stored. We talk about how they work, what they are used for, their advantages and disadvantages and more. We also look at how these technologies have changed over the past twenty years and point out the important developments and examples of when they have been used successfully. This analysis helps readers understand how well each technology works, its limits and its role in the progression toward a sustainable energy future. By exploring these technologies, we aim to provide a clear picture of how renewable energy sources and storage solutions can be used in microgrids to maximize energy independence.

Photovoltaic Solar Energy

Principle: Uses PV panels to directly convert sunlight into electricity through the photoelectric effect. When sunlight (photons) hits these cells, it causes electrons to be knocked loose from their atoms, generating a flow of electricity.

Applications: Power supply for homes, devices, and systems connected to grids or standalone setups (off-grid).

Advantages: Low environmental impact, adaptable installation, and decreasing costs.

Disadvantages: Dependent on sunlight and requires energy storage for continuous availability.

PV power is a technology that directly converts solar energy into electricity using PV cells. These cells function by absorbing photons from sunlight, which then generate an electric current through the PV effect. Over the past decade, this technology has experienced exponential growth due to advances in cell efficiency and significant reductions in production costs. Supportive policies, such as subsidies and tax incentives, have also played a crucial role in this expansion. Photovoltaic solar power has evolved to become a cost competitive energy source compared to fossil fuels. Innovations in materials and manufacturing techniques have increased the efficiency of solar cells and reduced their $/KWh cost. Additionally, the integration of solar energy into buildings (BIPV) and energy storage systems has facilitated its adoption.

Forecasts for the next twenty years anticipate a continued increase in solar production capacity, supported by investments in research and development. Advances in energy storage technologies, such as flow batteries and smart grid systems, will enable better integration of intermittent solar energy into the overall energy mix. Furthermore, innovation in solar cell materials and architectures could improve efficiency and reduce costs, making PV solar power even more accessible and competitive.

Solar Thermal Energy

Principle: Captures sunlight's heat using solar thermal panels to heat water or heat transfer fluids.

Applications: Domestic heating, hot water production, or industrial use.

Advantages: Highly efficient for direct heat without converting to electricity.

Disadvantages: Requires direct sunlight exposure and often needs thermal storage systems.

Solar thermal energy uses the sun's heat to produce energy. Unlike photovoltaic solar power, which generates electricity directly, solar thermal energy captures solar energy in the form of heat, often used for water heating or electricity production via steam turbines. In the past ten years, this technology has seen notable advancements in solar concentration systems and heat storage materials. Production costs have decreased, and efficiency has increased, making this technology more viable and accessible.

Over the next twenty years, trends indicate a significant expansion of solar thermal applications, particularly in industrial processes and green hydrogen production. Innovations in heat storage and concentration systems promise to further improve the efficiency and cost-effectiveness of solar thermal energy.

Biogas

Principle: Produced from the fermentation of organic matter (agricultural, industrial, or household waste) in the absence of oxygen (anaerobic digestion). It mainly contains methane.

Applications: Heat generators, electricity production, vehicle fuel.

Advantages: Recycles organic waste and reduces methane emissions.

Disadvantages: Depends on the availability of organic materials and requires appropriate infrastructure.

Biogas is produced by the decomposition of organic matter in the absence of oxygen. In the past decade, biogas production has benefited from numerous technological advancements, particularly in anaerobic digesters and gas purification.

Using agricultural and industrial waste to produce biogas has also helped reduce greenhouse gas emissions and promote a circular economy.

In the future, biogas is expected to play a crucial role in the energy transition, with developments anticipated in optimizing digestion processes and integrating biogas production with other renewable energy systems. Waste valorization initiatives and supportive policies will continue to stimulate this technology.

Biofuel

Principle: Derived from biomass (plants like corn, sugarcane, or vegetable oils) through chemical or biological processes.

Applications: Replacement for fossil fuels in combustion engines (gasoline, diesel, aviation).

Advantages: Less polluting than fossil fuels, renewable in theory.

Disadvantages: May compete with food crops and requires significant resources (water, arable land).

Biofuels are produced from plant or animal biomass. In the last decade, this technology has evolved with the rise of second-generation biofuels, which use non-food raw materials, thus reducing competition with food crops. Advances in conversion processes and biotechnologies have also improved efficiency and reduced production costs.

In the next twenty years, biofuels are expected to see increased adoption, particularly in the transportation and aviation sectors, where decarbonization options are limited. Research in biotechnology and process engineering promises to continue improving the sustainable production of biofuels.

Wind Energy

Principle: Converts wind's kinetic energy into electricity using wind turbines.

Applications: Wind power plants, hybrid systems (with solar or batteries).

Advantages: No direct greenhouse gas emissions, mature technology.

Disadvantages: Wind variability and potential visual and noise impacts on the environment. Additionally, the manufacturing and installation of wind turbines have a carbon footprint due to the energy and resources required for materials, production, transportation, and construction. Despite this, wind energy's overall carbon footprint is significantly lower compared to fossil fuels, with life-cycle emissions estimated at around 11 grams of CO_2 per KWh of electricity generated [50].

Wind energy converts the kinetic energy of the wind into electricity using turbines. The past decade has seen a dramatic increase in installed wind capacity, thanks to advances in turbine design, wind farm optimization, and cost reductions. Offshore wind farms, in particular, have grown rapidly due to their advantages of more consistent wind resources and less visual and noise impact.

Future trends for wind energy include continued expansion of offshore wind farms, improvement in materials and construction techniques, and the development of floating wind turbine technologies. These innovations will enable the exploitation of deeper marine areas and increase wind energy's contribution to the global energy mix.

Hydroelectric Power

Principle: Uses the movement of water (falls, currents, tides) to generate electricity by spinning turbines.

Applications: Hydropower dams, marine turbines for ocean currents.

Advantages: Stable and predictable source, high energy capacity.

Disadvantages: Can disrupt aquatic ecosystems and requires large infrastructure.

Hydroelectric power uses the force of water to generate electricity. This technology is one of the oldest and most established among renewable energies. The past decade has seen progress in small hydroelectric plants and systems for managing existing dams to minimize their environmental impact. Hydropower remains a reliable and sustainable source of renewable energy.

For the next twenty years, trends include improving water management systems, integrating hydropower with other renewable energy sources, and developing innovative technologies like low-impact hydropower. These developments will maximize the use of water resources while preserving ecosystems.

Geothermal Energy

Principle: Harnesses heat stored beneath the Earth's surface to produce electricity or heating.

Applications: Urban heating, electricity generation in geothermal zones.

Advantages: Very low CO_2 emissions, continuous availability.

Disadvantages: Limited to specific geographic areas and high initial costs.

Geothermal energy harnesses the Earth's heat to produce electricity and heating. Over the past decade, technological advances have enabled the exploitation of geothermal resources at greater depths and in previously inaccessible areas. Enhanced geothermal systems (EGS) and advanced drilling techniques have opened new possibilities for this renewable energy source.

In the future, geothermal energy is expected to play a more significant role in the global energy mix, with innovations anticipated in drilling technologies, integrating geothermal systems with other renewable energy sources, and expanding geothermal applications for urban and industrial heating.

Wave and Tidal Energy

Principle: Converts the mechanical energy of waves, tides, or currents into electricity using specialized devices.

Applications: Coastal electricity generation.

Advantages: High energy density and predictability of marine cycles.

Disadvantages: Still in development, potential impacts on marine life.

Tidal energy uses the movement of tides to generate electricity. This still-developing technology has seen significant progress over the past decade, particularly in energy capture devices and storage systems. Production costs remain high, but pilot projects and small-scale installations have demonstrated the potential of this renewable energy source.

Future trends for tidal energy include developing more efficient and cost-effective technologies, increasing installed capacities, and integrating with other marine energy sources, such as wave and current energy. These innovations will fully harness the ocean's energy potential.

Green Hydrogen

Principle: Produced by electrolyzing water using renewable energy to generate hydrogen, a clean fuel. It can also be produced by steam methane reformation (SMR) and gasification.

Applications: Energy storage, fuel for vehicles, and industrial uses.

Advantages: Zero emissions during use.

Disadvantages: Production process is still expensive, and infrastructure is limited.

Hydrogen is considered a key energy carrier for the transition to a decarbonized economy. The past decade has seen major advances in green hydrogen production, from renewable sources like wind and solar through water electrolysis. The reduction in costs and improvement in the efficiency of electrolyzers have facilitated the adoption of this technology.

Over the next twenty years, trends include the expansion of green hydrogen production, the development of storage and distribution infrastructures, and the integration of hydrogen into various industrial and transportation sectors. Technological innovations and supportive policies will play a crucial role in realizing hydrogen's potential as a clean energy source.

Wood and Solid Biomass

Principle: Combustion or gasification of solid organic materials (wood, agricultural waste) to produce heat or electricity.

Applications: Domestic heating, industrial boilers, electricity production.

Advantages: Widely available and carbon-neutral (if sustainably managed).

Disadvantages: Biomass combustion generates fine particle emissions and requires sustainable resource management. Gasification can generate toxic waste that requires careful disposal.

Biomass is a renewable energy source derived from organic materials such as agricultural residues, forest waste, and energy crops. The past decade has seen progress in converting biomass into electricity, heat, and biofuels. Cogeneration systems, which produce both electricity and heat, have particularly benefited from these advancements.

Future trends for biomass include optimizing conversion processes, improving energy yields, and reducing environmental impacts. Biomass will continue to play an important role in diversifying the energy mix, particularly in rural areas and energy-intensive industries.

Storage Solutions

Lithium-Ion Batteries (li-ion)

Principle: Lithium-ion batteries are rechargeable batteries that use lithium ions as the primary component of their electrochemistry. They are known for their high energy density, long cycle life, and low self-discharge rate.

Applications:

- Portable electronics (smartphones, laptops, tablets)
- Electric vehicles (EVs)
- Grid storage and renewable energy integration.

Advantages:

- High energy density
- Long cycle life
- Low self-discharge rate
- Wide range of applications.

Disadvantages:

- High initial cost
- Limited availability of raw materials (lithium, cobalt)
- Safety concerns (risk of overheating and fires)
- Environmental impact of mining and disposal.

Lithium-ion batteries are among the most prevalent energy storage technologies due to their high energy density, long cycle life, and efficiency. Over the past decade, significant advancements have been made in improving the performance and safety of these batteries, including the development of new materials such as lithium iron phosphate (LFP) and nickel-cobalt-manganese (NCM) chemistries.

These improvements have made Li-ion batteries more cost-effective and reliable, fostering their widespread adoption in various sectors, particularly in EVs and grid storage applications.

Looking ahead, the next twenty years are expected to bring further enhancements in lithium battery technology and manufacturing processes, leading to even higher energy densities, faster charging times, and reduced costs. Additionally, the recycling and reuse of lithium-ion batteries will become increasingly important to address environmental concerns and resource constraints.

Lead-acid Batteries

Principle: Lead-acid batteries are one of the oldest types of rechargeable batteries, using lead dioxide and sponge lead in an electrolyte of sulfuric acid.

Applications:

- Automotive applications (car batteries)
- Backup power supplies (uninterruptible power supplies, UPS)
- Large-scale energy storage systems.

Advantages:

- Low cost
- High reliability
- Mature and established technology
- Recyclability of components.

Disadvantages:

- Low energy density
- Shorter cycle life compared to other technologies
- Heavy and bulky
- Environmental concerns related to lead content.

Lead-acid batteries are one of the oldest and most established energy storage technologies. Over the past decade, enhancements in battery design and manufacturing have improved their efficiency, durability, and environmental impact. Innovations such as advanced lead-carbon (PbC) batteries have extended the cycle life and increased the power density, making them more suitable for applications requiring frequent cycling.

In the future, lead-acid batteries are expected to continue playing a crucial role in applications where cost-effectiveness and reliability are paramount, such as automotive starter batteries and backup power systems. Advances in recycling technologies and materials innovation will further mitigate the environmental impact associated with lead-acid batteries.

Sodium-sulfur Batteries (Na-S)

Principle: Sodium-sulfur batteries operate at high temperatures and use liquid sodium and sulfur to store energy, offering high energy density and efficiency.

Applications:

- Grid storage and renewable energy integration
- Large-scale energy storage systems
- Remote power applications.

Advantages:

- High energy density
- High efficiency
- Long cycle life
- Suitable for large-scale applications.

Disadvantages:

- High operating temperature
- Complex cooling and insulation requirements
- Safety concerns (corrosive materials).

Sodium-sulfur batteries, operating at high temperatures, have gained attention for their high energy density and efficiency. In the past decade, technological advancements have improved the safety, lifespan, and performance of these batteries, making them more viable for large-scale energy storage applications. Their use has expanded in grid storage and renewable energy integration, particularly in regions with high renewable penetration.

Over the next twenty years, sodium-sulfur batteries are expected to see increased deployment in grid-scale storage systems, driven by continuous improvements in materials and system designs. Research efforts will focus on enhancing the safety and reducing the operating temperatures, making them more competitive with other storage technologies.

Supercapacitors (Ultracapacitors)

Principle: Supercapacitors, also known as ultracapacitors, store energy through electrostatic mechanisms rather than chemical reactions, offering high power density and rapid charge/discharge rates.

Applications:

- Backup power supplies
- Electric vehicles (regenerative braking systems)
- Consumer electronics
- Grid frequency regulation.

Advantages:

- High power density
- Rapid charge and discharge rates
- Long cycle life
- Wide operating temperature range.

Disadvantages:

- Low energy density compared to batteries
- Higher cost per energy unit
- Limited energy storage capacity
- Requires complex control systems.

Supercapacitors, known for their high power density and rapid charge/discharge capabilities, have seen significant advancements in the past decade. Improvements in electrode materials, such as graphene and carbon nanotubes, have enhanced their performance and energy storage capacity. These advancements have expanded their applications in electric vehicles, renewable energy systems, and consumer electronics.

Future trends indicate that supercapacitors will continue to evolve, with research focused on increasing energy density and reducing costs. Hybrid systems combining supercapacitors with batteries are expected to become more prevalent, leveraging the strengths of both technologies for a wide range of applications.

Pumped Hydro Storage (PHS)

Principle: Pumped hydro storage involves pumping water to a higher elevation during periods of low energy demand and releasing it to generate electricity during peak demand, utilizing gravitational energy.

Applications:

- Large-scale energy storage
- Grid stabilization
- Renewable energy integration.

Advantages:

- High storage capacity
- Long operational life
- High efficiency
- Mature and reliable technology.

Disadvantages:

- High initial capital cost
- Environmental impact on aquatic ecosystems
- Geographical limitations (requires suitable terrain)
- Long construction times.

Pumped hydro storage remains the most mature and widely used large-scale energy storage technology. In the past decade, improvements in turbine and pump efficiency, as well as advancements in control systems, have enhanced the performance and flexibility of PHS systems. New projects have been developed, particularly in regions with abundant water resources and suitable topography.

Looking forward, PHS is expected to continue playing a vital role in grid stability and renewable energy integration. Innovations such as closed-loop systems and underground pumped hydro storage will expand the geographical possibilities and reduce environmental impacts. The continued development of smart grid technologies will also enhance the operational efficiency of PHS systems.

Compressed Air Energy Storage (CAES)

Principle: Compressed air energy storage systems store energy by compressing air and storing it in underground caverns, which is then released to drive turbines and generate electricity when needed.

Applications:

- Grid storage and renewable energy integration
- Peak shaving and load leveling
- Backup power supplies.

Advantages:

- Large storage capacity
- Long operational life
- High efficiency (with advanced systems)
- Low environmental impact (compared to other large-scale storage).

Disadvantages:

- High initial capital cost
- Requires specific geological conditions
- Efficiency loss during energy conversion
- Complex system design.

Compressed air energy storage has seen technological advancements in the past decade, particularly in the development of advanced adiabatic and isothermal CAES systems. These improvements have increased the efficiency and reduced the environmental impact of CAES. Pilot projects and commercial installations have demonstrated the potential of CAES for large-scale energy storage and grid stabilization.

In the next twenty years, CAES is expected to become more competitive with other storage technologies, driven by further advancements in compression and expansion technologies. The integration of CAES with renewable energy sources and smart grid systems will enhance its role in balancing supply and demand.

Flywheels

Principle: Flywheels store energy in the form of rotational kinetic energy, using a rotating mass (flywheel). They are used for short-term energy storage and grid frequency regulation.

Applications:

- Grid frequency regulation
- Uninterruptible power supplies (UPS)
- Transportation (regenerative braking systems)
- Renewable energy integration.

Advantages:

- High power density
- Rapid response time
- Long cycle life
- Low maintenance.

Disadvantages:

- Low energy density
- High initial cost
- Mechanical wear and tear
- Limited storage duration.

Flywheel energy storage systems have undergone significant advancements in the past decade, with improvements in materials, design, and control systems. Flywheels function by storing energy in the form of rotational kinetic energy. When energy is needed, the flywheel's rotational speed is converted back into electrical energy through a generator. Modern flywheels use high-strength composite materials and advanced magnetic bearings, increasing their energy storage capacity and efficiency. These advancements have expanded their applications in grid frequency regulation, uninterruptible power supplies, and transportation systems.

Future trends suggest that flywheels will continue to evolve, with research focused on increasing energy density and reducing costs. Integration with renewable energy systems and hybrid storage solutions will further enhance their role in the energy storage landscape.

Thermal Storage

Principle: Thermal storage systems store energy in the form of heat, which can be used for heating, cooling, or electricity generation. Examples include molten salt storage for concentrated solar power plants and phase change materials for building heating and cooling.

Applications:

- Concentrated solar power (CSP) plants
- Building heating and cooling
- Industrial process heat
- District heating systems.

Advantages:

- High storage efficiency
- Long storage duration
- Cost-effective (depending on materials used)
- Suitable for various applications.

Disadvantages:

- Thermal losses over time
- High initial cost (for some systems)
- Requires large storage volumes (for certain applications)
- Complexity in system design.

Thermal energy storage has seen notable advancements in the past decade, particularly in CSP plants and building heating and cooling systems. Innovations in heat storage materials, such as molten salts and phase change materials, have improved the efficiency and cost-effectiveness of thermal storage. These advancements have enabled more extensive deployment in industrial processes and district heating systems.

Looking ahead, thermal storage is expected to play a critical role in the energy transition, with continued research focused on improving material properties and system designs. Integration with other renewable energy sources and the development of large-scale thermal storage solutions will enhance the flexibility and reliability of energy systems.

Superconducting Magnetic Energy Storage (SMES)

Principle: SMES systems store energy in the magnetic field of a superconducting coil. They offer very high efficiency and rapid response times, making them suitable for applications requiring precise energy control.

Applications:

- Grid stabilization
- Voltage regulation
- Power quality improvement
- Research and scientific applications.

Advantages:

- Very high efficiency
- Rapid response time
- Long cycle life
- High power density.

Disadvantages:

- High initial cost
- Requires cryogenic cooling
- Limited energy storage capacity
- Technical complexity.

SMES have made significant progress in the past decade, with advancements in superconducting materials and cryogenic technologies. These improvements have increased the efficiency and reduced the cost of SMES, making them more viable for applications requiring rapid response and high power density. SMES has found applications in grid stabilization, voltage regulation, and power quality improvement.

In the next twenty years, SMES is expected to benefit from ongoing research in high-temperature superconductors and advanced cryogenic systems. These innovations will enhance the performance and cost-effectiveness of SMES, expanding its applications in the energy storage market.

Hydrogen Storage

Principle: Hydrogen storage involves storing energy in the form of hydrogen gas, which can be used for electricity generation, heating, or as a fuel for vehicles. It is a key technology for the transition to a decarbonized energy system.

Applications:

- Energy storage and grid balancing
- Transportation (fuel cell vehicles)
- Industrial processes
- Backup power supplies.

Advantages:

- High energy density
- Flexible applications (electricity, heat, fuel)
- Zero-emission when used in fuel cells
- Long-term storage capability.

Disadvantages:

- High initial cost
- Storage and transportation challenges
- Low overall efficiency (compared to direct electricity use)
- Requires significant infrastructure development.

Hydrogen storage has emerged as a key technology for the transition to a low-carbon energy system. Over the past decade, significant advancements have been made in hydrogen production, storage, and distribution technologies. In particular, the development of green hydrogen, produced from renewable energy sources via electrolysis, has gained momentum. Innovations in storage materials and compression technologies have improved the efficiency and safety of hydrogen storage.

Looking forward, hydrogen is expected to play a pivotal role in various sectors, including transportation, industrial processes, and power generation. Research efforts will focus on reducing the cost of hydrogen production and improving storage technologies. The development of hydrogen infrastructure and policies supporting hydrogen deployment will drive its adoption and integration into the energy system.

In the last twenty years, we have learned a lot about how to find and store renewable energy. This has shown us how these solutions can change our energy systems. Byunderstanding how each technology works, we can see how they can help us have energy that is independent and sustainable. As these technologies improve and mature, these systems will be able to be used more in microgrids. This will provide communities worldwide with energy that is clean, reliable, and strong.

Sizes and Capacities of Energy Sources

In this section, we describe some basic concepts and variables related to electrical power and energy and compare how much energy different sources can produce. We will compare nuclear power plants and solar panels, looking at the good and bad points of each one. This will help readers understand how much energy is needed to meet different demands and the implications of choosing one energy source over another. By understanding these concepts, readers will be better equipped to make informed decisions about energy solutions for their specific needs. The goal here is to provide a representation of the sizes and capacities of different energy sources.

Additionally, we will present data on electricity consumption by country and per capita indicators. This comparison is crucial for understanding the varying energy demands across different regions and how microgrids can be tailored to meet these specific needs. By analyzing electricity consumption data, readers can gain insights into the scale of energy usage and identify opportunities for implementing microgrids to enhance energy efficiency and resilience. For example, countries with high per capita electricity consumption may benefit from microgrids to manage peak loads and integrate renewable energy sources more effectively. Conversely, regions with lower consumption might use microgrids to improve access to reliable power and support local energy independence. Understanding these concepts equips readers to make informed decisions about energy solutions for their specific needs, highlighting the potential of microgrids to address diverse energy challenges globally.

Basic Concepts of Electrical Power and Energy

Electrical Power

Electrical power (P) is a measure of the amount of electrical energy transferred or converted per unit of time. It is expressed in watts (W), where one watt equals one joule per second (J/s). The general formula for calculating electrical power is given by:

$$P = VI$$

where:

- V is the electric voltage (or potential difference) in volts (V),
- I is the electric current in amperes (A).

Electrical power can also be expressed in terms of resistance (R) of a circuit using Ohm's law ($V = IR$), resulting in the following formulas:

$$P = I^2R$$

$$P = V^2/R$$

Electrical Energy

Electrical energy (E) is the capacity of an electrical system to perform work. It represents the accumulation or consumption of energy over a period of time. Electrical energy is measured in joules (J), but in the context of electricity consumption, it is often expressed in kilowatt-hours (kWh), where 1 kWh equals 3.6 million joules (3.6 MJ). The relationship between electrical energy, power, and time is given by:

$$E = Pt$$

where:

- E is the energy in joules (J) or kilowatt-hours (kWh),
- P is the power in watts (W) or kilowatts (kW),
- t is the time in seconds (s) or hours (h).

Thus, if an electrical device consumes 1 kW of power for 1 hour, it will have used 1 kWh of electrical energy.

Understanding these two concepts is essential for the analysis and design of electrical systems, as it allows for quantifying the efficiency and energy requirements of electrical devices. Additionally, these notions are fundamental for managing energy in electrical grids, billing for electricity consumption, and developing renewable energy technologies.

To understand how different electricity generation technologies compare, it's useful to compare a modern nuclear power plant with solar panel systems. Nuclear energy is rarely considered for microgrids, but this comparison shows the differences in power density, reliability, how much electricity can be produced, and environmental impact. This is important information before we look at solar systems, which are much more useful for home and local energy solutions.

Nuclear Power Plant Versus Solar Panels

Nuclear Power Plant: Key Numbers

A modern nuclear power plant is a large facility that generates electricity. It can produce an average of 1 GW, which is the same as 1,000 MW of continuous power. When working at full capacity all year round, a nuclear power plant can make about 8.76 TWh of electricity each year. This is enough to provide electricity for around 800,000 average households. The process of generating electricity in a nuclear power plant involves nuclear fission. This is a reaction where uranium atoms are split into smaller atoms, releasing a significant amount of heat. This heat is used to produce steam, which then drives turbines to generate electricity. The electricity produced is sent to the electrical grid, where it is distributed to homes, businesses, and industries. Nuclear power plants are known for their ability to produce large amounts of electricity without emitting greenhouse gases during operation, making them a cleaner alternative to fossil fuel-based power plants. However, they do require careful management of radioactive materials and have high initial construction costs.

Operating Principles of a Nuclear Power Plant

A nuclear power plant operates primarily on the principle of nuclear fission. This process begins in the reactor core, where uranium atoms are bombarded by neutrons. This reaction causes the uranium atoms to split into smaller atoms, releasing a substantial amount of energy in the form of heat. This heat is then used to heat water under high pressure in a primary circuit, which subsequently produces steam in a steam generator.

The generated steam is directed towards a turbine, where it drives the turbine blades, converting thermal energy into mechanical energy. This mechanical energy turns an alternator, which transforms the mechanical energy into electrical energy. This electricity is then transmitted to the electrical grid for distribution.

After passing through the turbine, the steam is condensed back into water by a cooling system and returned to the primary circuit for reuse. This cooling system typically uses water drawn from a natural source, such as a river or lake, to dissipate residual heat.

Efficiency and Electrical Notions

- **Power:** The power output of a nuclear power plant is often measured in MW or GW. For example, a typical nuclear power plant can have a production capacity of 1 GW, which corresponds to the amount of electricity needed to power approximately. This is enough to provide the yearly energy that around 800,000 average households need.
- **Energy Produced:** The energy produced by a nuclear power plant is measured in TWh. A 1 GW plant operating at full capacity for one year produces approximately 8.76 TWh of energy.

Examples of Nuclear Power Plants Worldwide

1. **Diablo Canyon Power Plant** (United States):
   - Location: California
   - Power Output: 2,256 MW (two reactors of 1,118 MW each)

- Annual Production: Approximately 18 TWh

2. **Gravelines Nuclear Power Station** (Europe, France):
    - Location: Northern France
    - Power Output: 5,460 MW (six reactors of 900 MW each)
    - Annual Production: Approximately 28.2 TWh
3. **Taishan Nuclear Power Plant** (China):
    - Location: Guangdong
    - Power Output: 3,500 MW (two reactors of 1,750 MW each)
    - Annual Production: Approximately 24.3 TWh

Advantages and Disadvantages of a Nuclear Power Plant

Nuclear power plants offer several advantages. Firstly, they provide stable and reliable energy production. Unlike renewable energy sources such as solar and wind, which depend on weather conditions, nuclear power plants can operate continuously and deliver a predictable amount of electricity. Additionally, nuclear power plants emit very low levels of greenhouse gases compared to fossil fuel-based thermal power plants. While nuclear reactors themselves do not produce carbon dioxide during operation, the processes involved in mining and refining uranium, manufacturing reactor fuel, and constructing the plants do result in some emissions. Furthermore, developing nuclear energy can strengthen a country's energy independence, reducing its reliance on energy imports.

However, there are also significant disadvantages. The management of nuclear waste is a major challenge, as it remains hazardous and radioactive for thousands of years, necessitating safe and durable storage solutions. The safety risks associated with nuclear power plants cannot be ignored. Although rare, nuclear accidents can have devastating consequences for the environment and human health, as demonstrated by the Chernobyl and Fukushima disasters. Finally, the costs of constructing, maintaining, and decommissioning nuclear power plants are very high, requiring substantial financial investments and extended construction timelines.

Solar Panels

Solar Panels: Key Numbers

- **Average Production:** The production of a solar panel depends on several factors, mostly related to the availability of sunlight, but on average a 1 m^2 panel produces about 150–250 W.
- **Required Area:** To reach a capacity of 1 GW with solar panels, about 4–7 km^2 of panels would be needed (depending on the efficiency of the panels).
- **Annual Production:** To replace the 8.76 TWh produced by a nuclear power plant, approximately 35–58 km^2 of solar panels would be required.

Operating Principles of a Solar Photovoltaic Power Plant

A solar PV power plant operates on the principle of the photovoltaic effect. This process begins with solar panels, which are composed of numerous PV cells made from semiconductor materials, primarily silicon. The generated DC electricity is then passed through an inverter, which converts it into AC electricity, suitable for use in homes and businesses. This AC electricity is then fed into the electrical grid for distribution.

Efficiency and Electrical Notions

- **Power Output:** The power output of a solar PV power plant is measured in MW or GW. For example, a large solar farm might be able to produce 100 MW of electricity, which is enough to supply around 12,500 US households each year. This number shows the typical amount of electricity that solar panels would produce if they were 20% efficient and if a typical household used 14,000 kWh of electricity per year.
- **Energy Produced:** The energy produced by a solar PV power plant is measured in MWh or GWh. A 100 MW plant operating at a realistic capacity factor of 20% would produce approximately 175.2 GWh of energy per year [51].

Examples of Solar Photovoltaic Power Plants EWorldwide

1. **Nellis Solar Power Plant** (United States):
   - Location: Nevada
   - Power Output: 72 MW
   - Annual Production: Approximately 32 GWh
2. **Serpa Solar Power Plant** (Portugal):
   - Location: Serpa
   - Power Output: 11 MW
   - Annual Production: Approximately 20 GWh
3. **Pirapora Solar Complex** (Brazil):
   - Location: Pirapora
   - Power Output: 321 MW
   - Annual Production: Approximately 700 GWh

Advantages and Disadvantages of a Solar Photovoltaic Power Plant

Solar PV power plants offer several advantages. First, they provide a renewable and sustainable source of energy. Unlike fossil fuels, sunlight is abundant and inexhaustible. Additionally, solar power plants produce no greenhouse gas emissions during operation, contributing to the reduction of carbon footprints and combating climate change. Solar power plants also require relatively low maintenance compared to other power generation methods.

However, there are some disadvantages. The initial cost of installing solar panels and related infrastructure can be high. Solar power generation is also dependent on weather conditions and the amount of daylight available, which can lead to variability in power output. Energy storage solutions, such as batteries, are often required to manage this variability and ensure a steady supply of electricity.

Comparison between Nuclear Energy and Solar Energy

Aspect	*Nuclear Energy*	*Solar Energy*
Source and Operation	Fission of uranium or plutonium atoms	Conversion of solar radiation into electricity via PV panels or thermal systems
Energy Efficiency	About 33–37%	About 15–22%
Emissions	Low greenhouse gas emissions during operation	Minimal during operation, but environmental impact during panel manufacturing and recycling
Environmental Impact	Significant impact from uranium mining and waste management	Lesser impact, but considerations during manufacturing and recycling
Waste	Radioactive waste requiring secure management for thousands of years	Panels with a lifespan of 20–30 years, recyclable
Safety	Risk of catastrophic events (Chernobyl, Fukushima)	Minimal risks, mainly mechani-cal and electrical
Initial Costs	Very high for construction and startup	Lower costs for installation, accessible for projects of all sizes
Operational Costs	Relatively low once operational	Low maintenance costs, but depend on location and solar conditions
Resource Availability	Uranium is a finite resource	Sunlight is a renewable and abundant source
Climate Dependence	Independent of weather conditions	Dependent on solar radiation, limited in low sunlight areas or at night

Number of Nuclear Power Plants in Operation and Replacement Calculation

According to the IAEA, as of the end of December 2023, there were **413 nuclear reactors in operation** worldwide across **31 countries**, with a total **net electrical capacity of 371.5 GW** [52].

To replace these plants, we would need a total capacity of 400,000 MW (or 400 GW).

To replace this capacity with solar panels, we need to consider the efficiency of solar panels. On average, each square meter of solar panel can produce about 200 W. Hence, to produce 1 MW, it would require about 5,000 m^2 (5,000 solar panels of 200 W each). To replace 400 GW, it would therefore require approximately 2,000 km^2 of solar panels.

Electricity Consumption by Country and Indicators Per Capita

Comparing different energy sources, like nuclear power plants and solar panels, shows the pros and cons of each option (Figure 2.8). To make the right choices about energy for different uses, we need to understand the basics of electricity and energy. This knowledge is essential for designing effective and sustainable energy systems that meet the needs of communities and industries. As we move towards a more decentralized and resilient energy future, these insights will guide us in selecting the right mix of energy sources to achieve our goals.

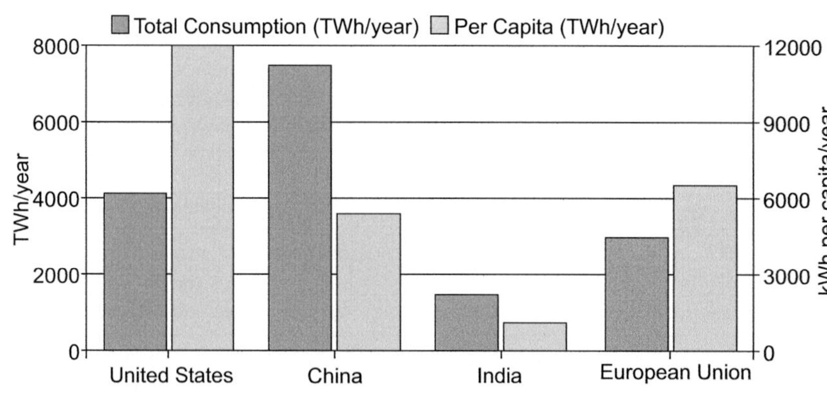

Fig. 2.8 Global electricity consumption and per capita indicators: A comparative analysis.

1. **United States:**
   - Total Consumption: Approximately 4,180 TWh per year.
   - Consumption per Capita: Approximately 12,000 kWh per capita per year.
2. **China:**
   - Total Consumption: Approximately 7,500 TWh per year.

- Consumption per Capita: Approximately 5,400 kWh per capita per year.

3. **India:**
   - Total Consumption: Approximately 1,500 TWh per year.
   - Consumption per Capita: Approximately 1,100 kWh per capita per year.
4. **European Union:**
   - Total Consumption: Approximately 3,000 TWh per year.
   - Consumption per Capita: Approximately 6,500 kWh per capita per year.

Conclusion: The Path to Energy Independence

As we conclude Chapter 2, it is evident that the journey towards energy independence is both complex and multifaceted. This chapter has provided a comprehensive overview of the building blocks necessary for achieving energy autonomy, emphasizing the importance of understanding and integrating various renewable energy sources, storage technologies, and smart grid systems. The chapter highlighted the significance of utilizing a mix of renewable energy sources, including solar, wind, hydroelectric, geothermal, and biomass. Each source has its unique advantages and challenges, and their combined use can enhance the reliability and sustainability of energy systems. Effective energy storage is crucial for balancing supply and demand, especially when dealing with intermittent renewable energy sources. Technologies such as lithium-ion batteries, flow batteries, thermal storage, and pumped hydro storage were discussed, each offering different benefits and applications.

The integration of smart grid technologies is essential for optimizing energy production, distribution, and consumption. These technologies enable real-time monitoring, demand response, and efficient energy management, contributing to the overall resilience and efficiency of the energy system. The chapter explored the differences between off-grid and on-grid systems, highlighting the benefits and challenges of each. Off-grid systems offer complete energy independence but require significant investment and maintenance. On-grid systems provide a reliable backup but may involve regulatory and interconnection complexities.

While the path to energy independence is promising, it is not without its challenges. Regulatory hurdles, technological limitations, and financial constraints can pose significant obstacles. However, these challenges also present opportunities for innovation, collaboration, and policy development. Navigating the complex regulatory landscape is crucial for the successful deployment of microgrids. Policymakers must also create supportive frameworks that encourage investment in new renewable, thermal and nuclear generation technologies (like high-efficiency solar power cells, small modular reactors, and fuel-flexible fuel cells and turbines) and facilitate the integration of decentralized energy systems. Continued research and development in energy technologies and energy storage solutions are essential.

Innovations in materials, efficiency, and scalability will also drive down costs and improve the performance of these systems. Developing and adopting new financial models that make renewable energy investments more accessible and attractive is vital. This includes exploring options such as power purchase agreements (PPAs), community solar programs, and green bonds. Raising awareness about the benefits of energy independence and educating the public on how to participate in and support these initiatives is critical. Community workshops, informational campaigns, and educational programs can play a significant role in fostering a culture of sustainability.

As we move forward, it is essential to remain adaptable and open to new ideas and technologies. The energy landscape is rapidly evolving, and staying informed about the latest developments will be key to achieving and maintaining energy independence. The journey towards energy autonomy is not just about technology and economics; it is also about people and communities working together to create a sustainable future. By embracing renewable energy, smart grid technologies, and innovative financial models, we can build resilient and self-sufficient energy systems that benefit everyone.

With a solid understanding of the building blocks of energy independence, we are now ready to delve into the practical aspects of implementing microgrids. Chapter 3, titled "The How: Engineering Your Energy Freedom", will guide you through the technical planning, project management, and regulatory considerations necessary for building a microgrid.

References

1. International Energy Agency. (2025). *Annual Change in Global Electricity Generation by Source, 2023–2024*. IEA. https://www.iea.org/data-and-statistics/charts/annual-change-in-global-electricity-generation-by-source-2023-2024.
2. Ember. (2024). *Global Electricity Review 2024*. Retrieved June 8, 2025, from https://ember-climate.org/insights/research/global-electricity-review-2024/.
3. https://www.woodmac.com/industry/power-and-renewables/us-solar-market-insight/.
   https://seia.org/research-resources/solar-market-insight-report-2024-year-in-review/.
4. https://www.solarpowereurope.org/press-releases/new-report-eu-solar-reaches-record-heights-of-56-gw-in-2023-but-warns-of-clouds-on-the-horizon.
5. Ember. (2023). *Global Electricity Review 2023*. Retrieved June 8, 2025, from https://ember-energy.org/latest-insights/global-electricity-review-2023/global-electricity-trends/
   Ember. (2024). *World Passes 30% Renewable Electricity Milestone*. Retrieved June 8, 2025, from https://ember-energy.org/latest-updates/world-passes-30-renewable-electricity-milestone/.
   Enerdata. (2024). *Renewables in Electricity Production. World Energy & Climate Statistics: Yearbook 2024*. Retrieved June 8, 2025, from https://yearbook.enerdata.net/renewables/renewable-in-electricity-production-share.html.
   McKinsey & Company. (2024). *Global Energy Perspective 2024*. Retrieved June 8, 2025, from https://www.mckinsey.com/industries/energy-and-materials/our-insights/global-energy-perspective.

Solar Business Hub. (2020). *Global Weighted-average Levelised Cost of Electricity of Utility-scale Solar PV Fell 82% between 2010 and 2019, IRENA Report finds*. Retrieved June 8, 2025, from https://solarbusinesshub.com/2020/06/06/global-weighted-average-levelised-cost-of-electricity-of-utility-scale-solar-pv-fell-82-between-2010-and-2019-irena-report-finds.

International Renewable Energy Agency (IRENA). (2019). *Renewable Power Generation Costs in 2018*. Retrieved June 8, 2025, from https://www.irena.org/Publications/2019/May/Renewable-power-generation-costs-in-2018.

Mercom India. (2023). *China's Solar Module Production Costs Plunge 42% in 2023*. Retrieved June 8, 2025, from https://mercomindia.com/chinas-solar-module-production-in-2023.

6. Wood Mackenzie. (2024). *US Microgrid Momentum: Energy Transition Applications in Action*. Retrieved June 8, 2025, from https://www.woodmac.com/news/opinion/ebook-us-microgrid-momentum-energy-transition-applications-in-action/.

   Center for Climate and Energy Solutions. (2023). *Microgrids*. Retrieved June 8, 2025, from https://www.c2es.org/content/microgrids/.
7. EnergySage – Refrigerator Usage: https://www.energysage.com/electricity/house-watts/how-many-watts-does-a-refrigerator-use/.
8. EIA – Lighting Consumption: https://www.eia.gov/todayinenergy/detail.php?id=38452
9. EnergySage – Laptop Usage: https://www.energysage.com/electricity/house-watts/how-many-watts-does-a-computer-use/.
10. Direct Energy – Dishwasher Usage: https://www.directenergy.com/learning-center/how-much-energy-dishwasher-use, EnergySage - Washing Machine Usage: https://www.energysage.com/electricity/house-watts/how-many-watts-does-a-washing-machine-use/
11. EnergySage – Refrigerator Usage: https://www.energysage.com/electricity/house-watts/how-many-watts-does-a-refrigerator-use/.
12. Eurostat – Lighting Consumption: https://ec.europa.eu/eurostat/statistics-explained/index.php/Energy_consumption_in_households&lang=en.
13. EcoCostSavings – Laptop Usage: https://ecocostsavings.com/how-many-watts-does-a-laptop-use/
14. *Source:* ElectricRate – Dishwasher Usage: https://www.electricrate.com/dishwasher-power-consumption/

    Source: In The Wash – Washing Machine Usage: https://inthewash.co.uk/washing-machines/how-much-electricity-does-a-washing-machine-use/
15. Energy.gov – HVAC Usage: https://www.energy.gov/energysaver/estimating-appliance-and-home-electronic-energy-use.
16. PCMag – Entertainment Devices: https://www.pcmag.com/how-to/power-hungry-pc-how-much-electricity-computer-consumes.
17. Eurostat – Heating Consumption: https://ec.europa.eu/eurostat/web/products-eurostat-news/-/ddn-20220617-1.
18. IEA – Renewable Energy: https://www.iea.org/reports/renewables-2022.
19. EIA – US Household Consumption: https://www.eia.gov/tools/faqs/faq.php?id=99&t=3.
20. UKPower – EU Household Consumption: https://www.ukpower.co.uk/home-energy/average-household-gas-and-electricity-usage.
21. European Commission – EPBD: https://energy.ec.europa.eu/topics/energy-efficiency/energy-efficient-buildings/energy-performance-buildings-directive_en.
22. (a) How Many Watts Does A Refrigerator Use? | EnergySage. https://www.energysage.com/electricity/house-watts/how-many-watts-does-a-refrigerator-use/

(b) Regional variations in household lighting consumption driven by bulb .... https://www.eia.gov/todayinenergy/detail.php?id=38452.
(c) How Many Watts Does a Computer Use? | EnergySage. https://www.energysage.com/electricity/house-watts/how-many-watts-does-a-computer-use/.
(d) How Many Watts Does A Laptop Use – 1,084 Studied | With Data. https://ecocostsavings.com/how-many-watts-does-a-laptop-use/.
(e) How Much Electricity Does My Dishwasher Use? | Direct Energy. https://www.directenergy.com/learning-center/how-much-energy-dishwasher-use.
(f) How Many Watts Does A Washing Machine Use? | EnergySage. https://www.energysage.com/electricity/house-watts/how-many-watts-does-a-washing-machine-use/.
(g) How much electricity does a fridge use? | Homebuilding. https://www.homebuilding.co.uk/advice/how-much-electricity-does-a-fridge-use.
(h) How Much Power a Fridge Uses - in Watts, Cost & kWh – Reduction Revolution. https://reductionrevolution.com.au/blogs/how-to/fridge-power-consumption.
(i) How much energy does a fridge consume? - Coolblue. https://www.coolblue.de/en/advice/energy-consumption-fridge.html.
(j) Lighting - IEA - International Energy Agency. https://www.iea.org/energy-system/buildings/lighting.
(k) Global electricity consumption in lighting in the Net Zero Scenario .... https://www.iea.org/data-and-statistics/charts/global-electricity-consumption-in-lighting-in-the-net-zero-scenario-2010-2030.
(l) Frequently Asked Questions (FAQs) - U.S. Energy Information .... https://www.eia.gov/tools/faqs/faq.php?id=99&t=3.
(m) How Much Electricity Does a Washing Machine Use? - In The Wash. https://inthewash.co.uk/washing-machines/how-much-electricity-does-a-washing-machine-use/.
(n) How Much Energy Does My Washing Machine Use? https://www.directenergy.com/learning-center/how-much-energy-washing-machine-use.
(o) Power-Hungry PC? Here's How Much Electricity Your Computer Consumes - PCMag. https://www.pcmag.com/how-to/power-hungry-pc-how-much-electricity-computer-consumes.
(p) How Much Power Does a Laptop Use? (Answered!) | DeviceTests. https://devicetests.com/how-much-power-does-a-laptop-use.
(q) How Many Watts Does a Laptop Use and How Much Does It Cost to Run? https://www.makeuseof.com/how-many-watts-laptop-use-how-much-cost-to-run/.
(r) Dishwasher Electricity Usage [Wattage & kWh]. https://www.electricrate.com/dishwasher-power-consumption/.
(s) Dishwasher Energy Calculator: Watts and kWh. https://www.energybot.com/energy-usage/dishwasher.html.
(t) Energy vs. Electricity: What's the difference? https://www.mcecleanenergy.org/mce-news/energy-vs-electricity-whats-the-difference/.
(u) Energy IQ: Energy, power and electricity: What's the difference? - Cummins. https://www.cummins.com/news/2019/08/13/energy-iq-energy-power-and-electricity-whats-difference.
(v) Electricity, Energy, and Power: What's the Difference?. https://www.kirbyelectric.com/electricity-energy-power-difference/.

23. https://www.pv-magazine.fr/2021/07/05/les-lecons-tirees-du-plus-ancien-systeme-photovoltaique-connecte-deurope/.
24. National Renewable Energy Laboratory. (2024). Fall 2024 Solar Industry Update (NREL/TP-7A40-92257). Retrieved June 8, 2025, from https://docs.nrel.gov/docs/fy25osti/92257.pdf.

International Renewable Energy Agency. (2019). *Renewable Power Generation Costs in 2018*. Retrieved June 8, 2025, from https://www.irena.org/Publications/2019/May/Renewable-power-generation-costs-in-2018.

25. https://www.greentechmedia.com/articles/read/u-s-microgrid-growth-beats-analyst-estimates-revised-2020-capacity-project.
26. https://www.globaldata.com/store/report/microgrids-market-analysis/.
27. https://www.eia.gov/outlooks/steo/report/natgas.php.
28. https://oilreviewmiddleeast.com/exploration-production/onshore-middle-east-cheapest-source-of-new-oil-production?utm_source=chatgpt.com.
29. https://www.iea.org/reports/key-world-energy-statistics-2021/final-consumption.
https://www.iea.org/reports/world-energy-outlook-2023.
https://www.iea.org/reports/key-world-energy-statistics-2020/final-consumption.
30. Viret, L. (author). *Evaarist SA*.
31. https://www.worldbank.org/en/news/press-release/2021/06/07/report-universal-access-to-sustainable-energy-will-remain-elusive-without-addressing-inequalities.
32. https://www.worldbank.org/en/news/press-release/2022/09/27/solar-mini-grids-could-power-half-a-billion-people-by-2030-if-action-is-taken-now.
33. World Bank. (2024, June 11). Progress on basic energy access reverses for first time in a decade [Press release]. Retrieved June 8, 2025, from https://www.worldbank.org/en/news/press-release/2024/06/11/progress-on-basic-energy-access-reverses-for-first-time-in-a-decade.
Sustainable Energy for All. (2024). *SEforALL Analysis of SDG7 Progress: 2024*. Retrieved June 8, 2025, from https://www.seforall.org/data-stories/seforall-analysis-of-sdg7-progress.
34. World Map with Percentages of Access to Electricity in 1990. (https://data.worldbank.org/indicator/EG.ELC.ACCS.ZS?end=2022&start=1990&view=map&year=1990)
35. World Map of countries with Access to Electricity in 2022 (in Percentages). (https://data.worldbank.org/indicator/EG.ELC.ACCS.ZS?end=2022&start=1990&view=map).
36. Source: Loïc Viret, Evaarist SA.
37. https://www.iea.org/reports/approximately-100-million-households-rely-on-rooftop-solar-pv-by-2030.
38. https://www.energy.gov/oe/microgrid-program-strategy.
https://www.energy.gov/oe/articles/two-new-reports-microgrid-rd-program-and-next-generation-grid-technologies.
39 *Source:* Loïc Viret, *Evaarist SA*.
40. Kanakadhurga, D., & Prabaharan, N. (2022). Demand side management in microgrid: A critical review of key issues and recent trends. *Renewable and Sustainable Energy Reviews*, *156*, 111915.
41. Balakumar, P. and Sathiya, S. (2017, April). Demand side management in smart grid using load shifting technique. In: *2017 IEEE International Conference on Electrical, Instrumentation, and Communication Engineering* (*ICEICE*), (pp. 1–6). IEEE.
42. https://www.eemg-mediators.eu/downloads/dynamic_pricing_in_electricity_supply-2017-2520-0003-01-e.pdf.
43. https://economics.mit.edu/sites/default/files/2022-09/Dynamic%20Pricing%20of%20Electricity%20%28with%20Catherine%20Wolfram%29.pdf.
44. https://op.europa.eu/en/publication-detail/-/publication/a8b8e55f-a17f-11eb-b85c-01aa75ed71a1/language-en.
45. https://link.springer.com/chapter/10.1007/978-1-4614-9035-7_3.
46. *Source:* https://www.congress.gov/bill/117th-congress/house-bill/5376/text
47. https://earth.org/electric-cars-environment/.

48. https://steerev.com/electric-vehicles-evs/environmental-benefits-of-electric-vehicles/.
49. https://www.energysage.com/electric-vehicles/energy-independence/.
50. https://www.factcheck.org/2018/03/wind-energys-carbon-footprint/. https://www.newscientist.com/lastword/mg24332461-400-what-is-the-carbon-payback-period-for-a-wind-turbine/.
51. https://shrinkthatfootprint.com/average-solar-energy-production/.
52. https://www.iaea.org/newscenter/news/iaea-releases-nuclear-power-data-and-operating-experience-for-2023?utm_source=chatgpt.com.

3 How Do We Implement a Microgrid?

Chapter 3 transitions from concept to execution by offering a practical roadmap for designing, financing, and deploying microgrids for various use cases. Starting with site assessment and load profiling, the chapter details the steps necessary to align energy generation and storage capacity with community or institutional requirements. The chapter examines microgrid architectures, from grid-connected to islandable systems, and discusses design flexibility based on geography, demand, and available renewable resources. The authors guide readers through key decisions, such as selecting generation technologies, integrating smart control systems, choosing battery types, and ensuring redundancy. The financial aspects are thoroughly addressed, including upfront costs, payback periods, and revenue models, such as net metering and grid services. The chapter compares different ownership models—utility-owned, third-party, cooperative, and community-driven—for their regulatory and financial implications. Readers are also introduced to incentive programs, tax credits, and policy frameworks that can reduce capital expenditures and accelerate returns on investment. Crucially, the chapter explores permitting processes, grid interconnection requirements, and cybersecurity considerations. The chapter provides guidance on engaging stakeholders, building public-private partnerships, and planning for scalability and maintenance. Real-world examples demonstrate how tailored approaches have succeeded in diverse environments, including rural villages, university campuses, and military bases. By the end, readers will be equipped with a strategic blueprint to initiate their own energy autonomy projects, armed with technical guidance and policy awareness.

You have identified your need or deep desire for energy autonomy and decided on the type of microgrid system to establish to achieve your goal. Now comes the hard part (yes, those countless meetings, e-mails, and Zoom/WhatsApp/Teams/plain-old telephone calls were the easy part). In this chapter, we focus on the technical planning, project management and regulatory hurdles you must overcome to establish your microgrid. You can use this chapter as the foundation for developing the checklists you will need to move forward with your plan. Spoiler alert—there will be a lot more emails, meetings, and calls as you work through how to implement your microgrid.

Before we dive into the considerations you need to make to plan the implementation of your microgrid, it is helpful to understand the bigger picture of the energy ecosystem which your actions will affect. Something a lot of people assume about interconnected grid systems and grids supplied by power from large

DOI: 10.1201/9781032689845-3

generation assets is that they have all kinds of redundant resilience systems already in place. It is almost unfathomable to most consumers that they could lose power for any reason beyond a transmission line failure. Surely, power can be drawn from other grids or the utility has DERs like numerous strategically placed renewable power charging battery storage backup systems or generators with automated controls that will route power where it is needed immediately. Right? Right…? (cue the peaceful background sound of crickets and frogs).

In reality, such systems and redundancies are quite rare (for now), and there are a host of reasons power outages can happen and often last a long time, even with no damage to transmission assets. We are not going to explore that universe of possible nightmare scenarios—that could easily be the subject of our next book—but instead the reasons why such systems are not commonplace. There are essentially four reasons this problem exists, and these same reasons will face you as you embark on our journey to autonomy: 1) the state, price and availability of desired technologies; 2) regulatory and legal restrictions, 3) availability of competent labor to install these technologies and, 4) available budget.

With the case of utility companies, one could posit that money is the biggest obstacle, followed by regulatory restrictions. In the case of public utilities that have monopolies on their service areas, they rely on getting approval from Public Utility Commissions (PUC) to charge ratepayers tariff surcharges to cover the cost of implementing large capital improvement projects, like the deployment of DERs, construction of large renewable or non-renewable power assets, new transmission lines, etc. PUCs have to weigh the obligation to ensure reliable power supply against the imposition of higher electricity rates to fund reliability enhancements for consumers [1] who are predictably and logically opposed to seeing their utility bills increase.

In the case of private utilities, their risk aversion and focus on dividend yields, share price, credit rating, etc., often prevents them from taking on the added cost of building in redundancies and resilience measures that do not generate regular, predictable revenues. They do not make money by reducing the probability of an outage by a decimal place or two but instead on selling power to their customers the 97–99.99% of the time it is available. The capitalist nature of these markets, which historically has passed the costs of reliability improvements along to the rate-paying customers, has been the largest impediment for utilities to build redundancy and resilience into the grid.

This is why we usually only see such improvements made when large sums of money (e.g., US Bipartisan Infrastructure Law) and tax incentives (e.g., US Inflation Reduction Act) are provided by the government(s) charged with overseeing the activities of these utilities, which can cover much of the capital costs of such improvements. Also, sometimes legislators step in and mandate certain changes to harden the power distribution system and increase reliability, which at least relieves the PUCs and affected utilities of the burden of balancing reliability and affordability but still ultimately imposes the cost on ratepayers [2]. Such initiatives, both direct government funding and legislative action to mandate reliability, can take a long time to draft and implement and can be frequently late to need for individual consumers in areas prone to outages or who are already weary of paying high rates for electricity. These delays, when combined with the steadily falling reliability of grid power in many locations, logically are driving more people to step forward and solve these problems themselves.

In your case, it seems you have already decided you want the security and freedom provided by energy autonomy, so money will be important but probably (hopefully) not the biggest obstacle you will encounter on your path. In what might seem a perverse turn of fate, some of your biggest obstacles will be (and, in many cases, already are) posed by the very institutions you are trying to put in the rearview mirror—utilities and the governments that regulate them, and sometimes even old laws particular to your jurisdictions. Let us explore this in more detail.

Concern #1: State and Availability of Technology

When it comes to what is available to build your microgrid, many people read about a fascinating technology related to renewable power generation, power control and distribution, or power storage, and assume it is widely available and affordable. The sad fact is that if you are reading or hearing about an amazing, new energy technology in a journal, newspaper, or from some other verifiable source (let us assume it is a legitimate report about a real technology), then the chances are it is something that is either still in prototype testing and not in commercial production, or it might have just begun commercial production. If the former, then you simply will not be able to buy it unless you are a billionaire. If the latter, then it will be almost surely prohibitively expensive to the point where only dedicated first-movers or institutions interested in testing its real-world efficacy would purchase it. Due to their need to keep costs reasonable and be able to plan a realistic project schedule, most communities, businesses, or institutions looking to build a microgrid will only consider mature, widely available, and therefore affordable technologies to be part of the system. So microgrid planners must first temper their expectations and design their system with these limitations in mind.

For example, the promise of small modular nuclear reactors ("SMR") with nameplate output in the 100KW–5MW range is tantalizing, and many articles have been written on them in a variety of technical journals and mainstream media. While the amount of press attention this technology has received might lead one to think SMRs are just rolling off of assembly lines, none that would be suitable for a community, business, or installation seeking energy autonomy are currently commercially producing power or are in commercial production. Only recently has an SMR developed by Nuscale received design approval from the US Nuclear Regulatory Commission, and its reactor has a nameplate output of 77 MW [3], which is good for large-scale industrial applications, data centers and large municipalities but not most of the microgrid deployment types we contemplate in this book.

However, photovoltaic solar, small-scale geothermal, micro hydroelectric turbines, medium-scale battery energy storage, pumped-energy storage, and bio-waste fuel powered generators and fuel cells are in commercial production and are affordable in many places (i.e., competitive with the current cost of electricity provided by utilities) or are becoming increasingly so. In other words, focus on what is and not on what you wish there was. Do a lot of research into what is available in the marketplace and from whom. A good rule of thumb (and coincidentally a US federal government procurement policy for certain types of purchases) is to ensure that what you want to buy can be provided by at least three different manufacturers or suppliers, so you can compare prices, capabilities, warranty terms, service costs, spare parts availability, etc.

In Europe, the state and availability of technology is comparable to what is available in the US. Similar to the US, the microgrid market remains relatively nascent. However, the more widespread proliferation of renewable power, especially with solar in Germany and wind power in Scandinavia, opportunities for and interest in microgrids are beginning to expand. Nevertheless, islandable microgrid projects in the 100 KW–5 MW range remain relatively rare.

Cost Figures for Photovoltaic Panels (PV)

One of the biggest factors influencing the affordability and availability of equipment for microgrid systems is the cost of purchasing and installing PV solar panels. The cost of a single 400-watt solar panel typically ranges between $400 and $600, depending on the purchasing method. Premium monocrystalline solar panels currently cost between $1 and $1.50 per watt. This information is based on the data provided by Solar.com in 2024 [4]. Solar energy is intermittent because it depends on sunlight, which is unpredictable in many parts of the world. Solar forecasts can be accurate only up to a few days in advance. Weather and installation data can help, but no prediction is perfect.

US Residential Solar Installation Costs

- **Average Cost per Watt (installed):** Between **$2.50 and $3.50**, encompassing panels, inverter, mounting hardware, wiring, and professional installation.
- **Typical System Size:** Residential systems generally range from **5 to 10 kilowatts (kW)**, leading to total installation costs of approximately **$12,500 to $35,000** before incentives.
- **Battery Storage:** Adding a battery system (e.g., Tesla Powerwall) can increase costs by **$10,000 to $15,000**.

European Market Comparison

- **Average Installed Cost:** Typically between **€1.20 and €2.50 per watt**, depending on country and system size [5].

- **Price Comparison:** While per-watt costs can be **lower in some European countries** (e.g., Germany or Spain), the **overall cost of a full residential installation** is often **comparable to or slightly higher (10–25%)** than in the US, especially in countries with higher labor or compliance costs.
- **Incentives:** Many EU countries offer subsidies, feed-in tariffs, or VAT reductions, which can significantly reduce net costs.

Economic Calculations

Solar panels can save you money in the long run because they mean you're less dependent on the power grid and you'll pay less for electricity. In some places, extra energy can be sold back to the grid. One way to see how much money you can save on electricity bills is to look at the payback period. This is the number of years it takes for the savings to cover the initial installation cost. If it takes less time to pay back, it's a quicker return on investment.

For example, let's think about how long it would take to make up for the cost of installing solar panels on a house. The total cost up front is $20,000. After installation, homeowners can save an average of $2,500 a year on electricity bills by relying less on the grid and generating solar energy.

To work out the payback period, divide the total upfront cost ($20,000) by the yearly savings on electricity bills ($2,500):

Payback Period = Total Upfront Cost/Annual Savings on Electricity Bills

Payback Period = $20,000/$2,500 = 8 years.

This means that the money you spend on the system would be returned through energy savings after about eight years. If you can get your money back in five or six years, that's a good sign. You'll get your money back more quickly and could end up saving more overall.

When comparing the cost of solar panels to grid electricity, it is important to consider two factors:

1. **Cost of Grid Electricity:** Compare the total cost of installing and maintaining solar panels with the expected cost of electricity from the grid over the system's lifespan (typically 25–30 years).
2. **Rate of Grid Electricity Price Increases:** Consider the historical rate of increase in grid electricity costs.

This comparison should be made objectively, without bias towards either option. Solar panels offer a fixed cost for the electricity they generate, protecting against future price increases.

The cost trends of commercial solar systems in the United States from 2010 to 2023 reveal a significant decreasing trend in costs. According to the Solar Energy Industries Association (SEIA) and Wood Mackenzie [6], the cost of commercial solar systems was approximately $5.92 per watt in 2010. This figure steadily declined, reaching $4.37 per watt in 2012 and further dropping to $2.42 per watt in 2015. The downward trend continued, with costs reducing to $1.83 per watt in 2018. By 2023, the cost of commercial solar systems had decreased to $1.37

per watt. This substantial reduction in costs can be attributed to advancements in solar technology, economies of scale, and increased market competition. These trends underscore the rapid progress in making solar energy more affordable and accessible, which is crucial for the widespread adoption of renewable energy sources.

Solar panels often save money and make you more energy independent. How long it takes to get your money back depends on things like where you are, any help from the government, and how much electricity you use. Think carefully about these factors before making a decision.

According to S&P Global Commodity Insights [7], the price of wind and solar PPAs in Europe from 2023 to 2024 showed notable fluctuations. In the first quarter of 2023, solar PPAs were priced at 73.2 euros MWh, while wind PPAs were at 106.06 euros MWh. Over the next few quarters, these prices varied, and by the first quarter of 2024, solar PPAs had decreased to 67.63 euros MWh, r and wind PPAs to 81.13 euros MWh. These trends reflect the dynamic nature of the renewable energy market in Europe, influenced by factors such as project costs, revenue forecasts, and market conditions.

As an example of how solar power systems are becoming less expensive and thus more popular in the United Kingdom, we turn to a data analysis based on recent insights from the Microgeneration Certification Scheme and the Energy Saving Trust [8],

Between 2016 and 2024, the cost of small-scale solar PV installations in the United Kingdom declined significantly, reflecting improved solar technologies, market competition, and increased supply chain efficiency. According to data from the Microgeneration Certification Scheme (MCS), the average cost of home solar installations fell from £9,238 in January 2024 to £7,561 in December, with the yearly average settling at £8,198. This represents an 18% drop from the 2023 average of £9,994 and marks a faster pace of decline than in previous years.

The Energy Saving Trust notes that a standard 3.5 kWp domestic system—around 10 panels—now costs roughly £7,000, making solar increasingly affordable for UK households. These reductions reflect advances in photovoltaic cell efficiency, simplified installation, and lower balance-of-system costs. While the number of certified installations in 2024 slightly declined to 184,462 (down 7% from 2023), cost trends continue downward.

Long-term comparisons show dramatic progress: from around £2,500 per kilowatt in 2016 to approximately £2,000–£2,500 per kilowatt today. Regional data highlight adoption patterns, with Anglesey leading at 3.18% of households equipped with solar PV, and remote areas like the Isles of Scilly and Shetland Islands showing strong uptake of battery storage technologies due to higher energy costs and reliability concerns.

These trends underline the role of solar PV in the UK's push toward net-zero emissions by 2050, bolstered by incentives such as the Boiler Upgrade Scheme and potential mandates for rooftop solar on new homes.

Concern #2: Who's Gonna Get in My Way?

The answer—a lot of people. Many of those very utilities mentioned before as not having the ability or desire to invest in the sort of resilience infrastructure needed to reduce the risk of outages also view the project you are planning as a considerable business threat based on the prospect of many similar projects breaking ground in their service area. Many people find this mind-boggling. Why would my somnolescent utility company care if I became more independent or even was able to start selling them excess power? Would it not help them reduce costs and give them a new source of power they did not even have to spend money to install? It seems like a win-win, right? Not for many.

As we mentioned before, one thing almost all utilities have in common is that they are very risk averse. They want to provide their investors with predictable, if modest, returns and dividends. They achieve this by not taking any unnecessary risks and by identifying and mitigating risks that manifest to this business model as quickly and resolutely as possible. Institutional investors and major fund managers buy utility shares as a low-risk anchor for their portfolios and vote in directors onto their boards to ensure the utilities deliver those predictable results and dividends [9]. Consequently, many utilities view this as a zero sum game. It is not their fault, but more a symptom of the market dynamics to which they are subjected and the regulatory environment in which they operate. If you want to get political about this, do not bother writing letters to the utility management asking for change, write letters to your elected officials and lobby for legislation that would facilitate the development and deployment of privately owned and operated, grid-interconnected, renewable energy assets, and microgrids. Be prepared to face some stiff headwinds though.

Initially, utilities and their related regulators offered little to no opposition to individual consumers installing their own generation systems and utilizing net metering regimes to sell excess power to the grid. However, as the costs to obtain and install these systems fell, incentives to do so abounded and people started adopting them with increasing frequency and speed, utilities and regulators began to take note and, in some cases, take steps to reduce the financial benefits of net metering to create less incentive for ratepayers to install these systems.

You present the risk to utility investors of depriving them of paying customers and introducing uncertainty into their financial and capital expenditure planning. You also present the potential for a paradigm shift. Most utilities are very vocal about claiming how they are investing in being more reliable, more climate-friendly, etc., but this is currently conditioned primarily on their ability to maintain the current paradigm of being your primary supplier of power and providing that power from large, centralized generation assets financed mostly by their ratepayers. A significant wave of communities and commercial or industrial customers moving toward the autonomic model would threaten the viability of this business model. The elimination of revenues from newly autonomous consumers combined with the requirement to pay them for excess power produced not only impacts their financial results negatively each quarter but also shifts the cost burden of maintaining the utilities' infrastructure and generation assets to the shrinking number of ratepayers who still depend on the utilities for power supply. This will increase those non-autonomous ratepayers' electricity bills and potentially create tension between the "haves" and "have-nots" of the energy marketplace.

Despite these emerging market and political dynamics, some utilities and state regulators are embracing the gradual paradigm shift toward decentralized power systems. These utilities are examining opportunities to work with communities to help establish their autonomy while being able to leverage the microgrids established in those communities or aggregated from multiple private behind-the-meter installations to add resilience and capacity to the larger grid. Find out which kind of utility market you live in and your utility's (and regulator's, if applicable) policies related to on-site customer power generation. If you unfortunately live in jurisdictions where the utilities and/or regulators are still embracing policies that discourage energy autonomy, then you need to plot your course on this journey very carefully. If you live in a more "enlightened" (pun, as usual, intended) jurisdiction, then make sure you take advantage of every incentive that is offered and to involve your utility and local government in your planning to maximize the benefits. Some of the more progressive utilities are examining the establishment of "virtual power plants" which would integrate private DERs into their existing infrastructure to optimize grid performance, improve power quality and enhance resilience. Hopefully, more utilities will embrace this emerging paradigm that accepts and even welcomes new privately-owned DERs, but for now it remains the exception, not the rule for them to do so.

It is outside of the scope of this book for us to attempt to identify every possible law in every jurisdiction that has the potential to inhibit or enable your work. A couple resources we found very informative while preparing this book highlight the complexity of these issues, the volatile, often conflicting, market dynamics and the quilt work state of the regulatory environment [10]. If we were to try to include a compendium of these laws in this book, we would exceed our word limit before we would be even 3% finished. These laws, regulations, ordinances, etc., vary from city to city, county to county, state to state, region to region, and from country to country and, in some cases, may even cross national frontiers where there are international grid interconnections. They are also in flux, with a myriad of people, companies, and organizations with vested interests and a lot of money to gain or lose at stake. Instead, we consider it more worthwhile to more generally familiarize you with certain terms and concerns you will encounter on this journey.

Interconnected

When considering the greater energy ecosystem in which you are planning to participate, let us start with the concept of "interconnection". This, as the literal meaning of the word implies, means the physical and electronic connection between your microgrid and the utility grid whence you have been drawing your power until now. Assuming you, as most people, wish to maintain that connection and essentially have the grid available as a backup source of power and also potentially as a source of revenue for surplus power your microgrid may produce, you will need to have some form of interconnect authorization or agreement. A good primer on what these are and the various forms they can take in the US is provided online by the US Department of Energy's National Renewable Energy Laboratory [11]. If your system is on the smaller side (generation capacities vary by jurisdiction, but usually anything under a megawatt qualifies), then you would sign up with your utility for "net metering" if it is allowed in your jurisdiction and by your utility.

In Front or Behind the Meter?

Other concepts like "in front of the meter" and "behind the meter" will come into play. Again, the literal meanings help explain these concepts and they would only apply if interconnection is contemplated. In many jurisdictions, there is a limit on how large of a generation asset can be placed "behind the meter" (i.e., on your property, behind your electricity meter, with net metering). If your microgrid's generation capacity exceeds that, it will have to be "in front of the meter" and interconnected into the grid as a large, new generation asset. This will require the execution of an interconnect agreement with the utility or regional distribution network this new asset would be supplying. It does not matter if most of the generation's supply is servicing your electrical load. In this instance, "size [only] matters". And the applicable size varies among jurisdictions. It can take six months to five years to execute these agreements and get authorization to connect your new asset to the grid. Also, as mentioned previously, many power-generating utilities are not excited about the prospect of an unaffiliated newcomer arriving with a new generation asset the utility does not own and which is reducing the KWh they had been selling previously to your community, facility, or business. Those utilities have little to no incentive to expedite the processing of your interconnection agreement and may look for ways to complicate its execution.

For a more detailed look into the US regulatory environment related to renewable power and energy efficiency, we turn to a data analysis based on the Database of State Incentives for Renewables & Efficiency (DSIRE) [12].

The landscape of state-level renewable energy and energy efficiency policy in the United States reveals wide variation in commitment to clean energy transition. Some states stand out as policy innovators, with robust portfolios of incentives, regulations, and support mechanisms that promote renewable deployment and improved energy efficiency across residential, commercial, and industrial sectors. As of 2025, California leads nationally with 156 total policies and incentives, reflecting its ambitious climate goals and comprehensive regulatory approach—including tax credits, rebate programs, building energy codes, and renewable portfolio standards.

Minnesota follows with 128 policies and incentives, demonstrating strong commitment to wind power and energy efficiency. Colorado ranks third with 113, closely followed by Texas with 112, showing how diverse energy economies are also advancing renewables. Other top-ranking states include Oregon (101), New York (93), Massachusetts (87), Washington (83), Florida (77), and Maryland (72), forming a geographically and politically diverse coalition united by their support for clean energy policy frameworks.

These leading states employ multifaceted strategies that include rebate programs (1,144 nationwide), loan initiatives (170), public building energy standards (96), property tax incentives (87), net metering (72), and solar/wind access laws (69), along with 49 states adopting some form of renewable portfolio standard. In contrast, several states offer fewer than 20 such incentives, illustrating the disparity in political will, resource availability, and policy orientation. Nonetheless, the spread of state-level innovation offers significant opportunities for policy learning and transfer. As federal frameworks evolve, state action remains a key driver in advancing the US clean energy transition and maintaining leadership in the global low-carbon economy.

Net Metering and RECs

Net metering allows you to sell back to the utility any excess power you produce. How much money you can recover (if any) for the excess power your system generates depends primarily on where you live, what the laws of your jurisdiction allow and what your utilities allow. The earlier cited National Academy of Sciences report on net metering provides an excellent and substantive review of the state of net metering in the US as of 2023 [13]. The report highlights the current financial benefits of net metering across US states as of 2023 [14]. This regulatory environment is changing so rapidly that the net metering benefits available as depicted on that map may (probably will) have changed by the time this book is published.

In some places, you may also have the option to produce renewable energy credits (RECs) [15] for excess power transmitted to the grid instead of net metering (assuming your system is based on renewable power). The value of those RECs fluctuate as any market-traded financial asset does. For example, this choice is available to Dominion Power customers in the US State of Virginia [16].

In short, if you wish to establish an interconnected microgrid larger than a single-home system, we strongly advise you retain the assistance of legal counsel that is well-versed in the power sector and related construction permitting laws, regulations, and ordinances in the applicable jurisdictions. They will be able to help you navigate these treacherous and unpredictable waters to ensure you achieve your goal without being frustrated, stymied, or defeated by regulators and possibly utility companies.

To examine the increase in the number of prosumers benefiting from net metering, we turn to a data analysis based on the US Energy Information Administration's Annual Electric Power Industry Report [17].

Net metering adoption in the US has grown substantially over the past decade, with participating customers increasing from approximately 156,000 in 2010 to over 4.1 million by 2022. This more than 26-fold rise reflects the rapid expansion of distributed solar photovoltaic systems across residential, commercial, and industrial sectors. Solar PV remains the dominant technology in net metering programs, though wind and battery storage systems paired with solar are gradually gaining traction.

This surge in participation highlights how American consumers are increasingly turning to distributed renewable energy to lower electricity bills, enhance energy independence, and contribute to climate goals. The US Energy Information Administration (EIA) collects and publishes detailed net metering data by technology type, customer class, state, and balancing authority, offering crucial insights into the ongoing transformation of the nation's electricity landscape through decentralized energy production.

These regulatory burdens and being subject to a regulatory and market environment that can be gently described as "in flux", in addition to the general desire to be less dependent on a less dependable grid, are the main reasons some people have decided to literally "cut the cord", i.e., go off-grid and become completely autonomous. In this instance, you might have to pay some service or penalty fee to disconnect from the grid. The only laws that would apply to your projects would be of the more mundane variety, namely the laws and ordinances your city, county, state/province and/or country might require for construction, air quality permitting (if your system would have any emissions) and anything else that might apply based on your system design. There is something else to consider though—certain laws in certain jurisdictions may actually make full grid disconnection for residences illegal and deem the home(s) legally unfit for habitation. For example, in Florida, as in many jurisdictions, the International Property Maintenance Code [18] has been instantiated into law. Judges have interpreted this Code to mandate hookups to utilities, especially in urban and suburban areas. The reasoning used is that such residences cannot comply with the minimum standards for safety and habitability without connection to utilities. In rural areas, there is much more leniency regarding application of this Code.

In Europe, consumers are more likely to buy electricity generated by microgrids, even at higher prices set by the main grid. Generally, the situation is similar to that in the US except in a few aspects. There are plenty of constraints that hinder these kinds of microgrid initiatives.

Like in the US, the EU regulatory frameworks are still evolving. This lack of a clear legal structure can make transactions more expensive and uncertain for microgrid developers. The European Commission has funded more than 80 microgrid projects under different programs. The Horizon Europe program has also funded clean energy solutions. There are several existing and developing microgrids across Europe. For instance, a renewable energy system at a wood pulp mill in Finland could produce an additional 2 TWh of renewable electricity. Community microgrids are becoming more popular in Europe. The European microgrid market is indeed expected to grow rapidly, with an estimated 10.65% annual growth from 2022 to 2030. This growth is driven by renewable energy sources and the need for resilient energy systems [19]. In Europe, more people are getting involved in renewable energy. Over 2,250 energy communities are already helping 1.5 million people produce electricity. By 2050, it is likely that 83% of EU citizens will be involved in similar projects [20].

However, there are many challenges to developing microgrids in Europe [21]. These include regulations, technical issues, and the need for a lot of money upfront. The costs of the current legal framework can be more than the benefits, making things more difficult. Despite these challenges, microgrids have a promising future in Europe. New technology and more collaboration could help Europe use microgrids to build a better, more secure, and more sustainable energy future [22].

Regarding the regulatory environment, in the EU, it is still evolving much like it is in the US. This lack of a clear legal structure can increase transaction costs and create uncertainty for microgrid development. We can also add that community microgrids are growing in popularity in Europe. These microgrids are clusters of interconnected energy resources that meet the energy needs of localized areas and regions [23]. However, despite the potential benefits, there are barriers to the development of microgrids in Europe. These include regulatory hurdles, technical challenges, and the need for significant initial investment [24]. All in all, compared to the US, Europe seems to have a more favorable attitude toward microgrid-generated electricity. Despite the obstacles, there are numerous successful microgrid implementations across Europe, and the future of microgrids in the region appears promising.

Concern #3: Who's Going to Build It and With What?

Most people do not think much about macroeconomics when they are planning a home improvement project or other small-scale construction effort. Availability of reasonably priced labor and materials is usually taken for granted, so procurement is typically not considered a significant risk factor. However, anyone who tried to get their home remodeled during the COVID pandemic learned very quickly about what happens when demand greatly outstrips supply. The cost of building materials and labor soared, and many contractors would simply not bid on new work or would walk away in the middle of a smaller job to take on a bigger one that offers the promise of higher profit margins.

The same trend has been developing in the clean energy sector as technologies become more mature and available as production scales up, and incentives for installing clean power abound. The passage of the Inflation Reduction Act turbocharged this marketplace in the US. It has become often challenging for even large entities like utilities and federal agencies to source enough of the right kind of competent labor and materials to begin planned projects. Labor pressures are reshaping supply chains, and businesses must navigate these challenges to ensure efficient operations and resilience in the face of disruptions [25].

To examine global trends in clean energy workforce development, we turn to a data analysis based on a review by the International Labour Organization and the International Renewable Energy Agency [26].

Global renewable energy employment reached a record 16.2 million jobs in 2023, reflecting an 18% increase from 13.7 million in 2022. China accounted for 7.4 million of these jobs—46% of the global total—followed by the European Union with 1.8 million, Brazil with 1.56 million, and both the US and India with close to one million each.

Solar PV technology led job creation, providing 7.2 million positions worldwide. Liquid biofuels supported 2.8 million jobs, with Brazil contributing about one-third of that figure. Hydropower accounted for 2.3 million jobs, and wind energy created 1.5 million. These figures underscore both the accelerating momentum of the global energy transition and the persistent regional disparities, as Africa, despite its abundant renewable resources, accounted for just 324,000 jobs in the sector.

Energy efficiency jobs reduce the amount of energy needed to make products and provide services. This includes jobs related to making buildings, industrial processes, and transportation systems more efficient. The IEA says that energy efficiency jobs often require a lot of labor [27]. They include retrofitting buildings, manufacturing efficient appliances, upgrading industrial equipment, promoting energy-saving practices, and implementing energy-saving measures in various sectors. Solar jobs are related to making, installing, and maintaining solar energy systems. This includes roles like manufacturing solar panels, marketing and installing solar power systems, and maintaining and repairing solar installations.

As an example of a more detailed breakdown of renewable energy and energy efficiency employment trends, we present a data analysis based on the latest employment figures from the UK's Office for National Statistics [28].

The UK's Low Carbon and Renewable Energy Economy (LCREE) employed 272,400 full-time equivalent (FTE) workers in 2022, marking an 8.0% increase from 252,300 in 2021. The energy efficient products group accounted for the largest share with 134,900 FTEs (49.5% of the total), followed by low carbon electricity (41,800 FTEs), low emission vehicles and infrastructure (34,800 FTEs), and smaller contributions from nuclear power, energy monitoring systems, and other renewable electricity activities.

By industry classification, construction led LCREE employment with 99,100 FTEs (36.4%), followed by manufacturing with 75,000 and professional, scientific, and technical activities with 46,200. This distribution highlights the UK's increasing focus on net-zero targets and demonstrates the importance of energy efficiency and low-emission transport as emerging drivers of green employment, alongside the continued relevance of traditional sectors such as construction and manufacturing.

These forces [29] will apply to most smaller project developers as well, possibly even more acutely than to large-scale developers. Some of these supply issues will take a long time to resolve, as more people are trained to install microgrids and clean energy systems and enter the workforce. Also, domestic sourcing limitations or import tariffs may crimp available supply of materials until domestic manufacturing ramps up, which can take years. Add to this the prospect of large project developers tying up available resources for long periods of time, and smaller microgrid developers are going to find procuring the work and materials quite challenging.

The only way to address this problem while the supply/demand equation slowly balances out is to plan ahead, be mentally prepared for a long procurement effort and begin engaging potential contractors early and often to gauge availability and support. It will also be necessary to do due diligence on those contractors to reduce your risk or performance issues after the project has begun. Check their and their owners' (if it is a small business) legal history (any open lawsuits, liens, bankruptcy, etc.). Make sure they and any subcontractors they plan to engage have all professional licenses and certifications required to do every aspect of the work and are properly insured. If the project is on the larger or more expensive side, then you may wish to ask about whether they would be able to provide a construction bond to ensure adequate funds will be available to finish the work if they default for whatever reason.

Finally, follow the "rule of three" we mentioned above—get at least three quotes from companies that you have found to be responsible and capable so you can compare prices, schedules, terms and conditions, degree and complexity of any subcontracting contemplated, recommended materials, and whether they plan to procure the more expensive materials (usually with a 5–15% markup) or give you the option to do so. Typically, the contractor should provide all construction materials and procure those at a time and in quantities needed to complete the project. You should not have to do that unless you want to self-perform a substantial part of project management. However, for the more expensive and technically complex materials, you may wish to consider procuring yourself to have full control over and insight into the delivery timelines and terms and not pay the markups. Be careful though, if there are any delays due to your involvement, your contractor will rightfully expect you to pay for any resulting costs.

As with evaluating your regulatory environment, our same advice goes for doing due diligence on potential contractors, preparing your request for proposals and negotiating the terms of the contract—hire a good lawyer that is experienced in this space. There is a high probability that the same lawyer that did the regulatory risk review for you can also help with the due diligence, or someone else in the lawyer's firm could. Legal firms like to offer turnkey solutions for their clients so will very likely have a team of lawyers experienced in the various aspects of the project you are considering. Fortunately (or unfortunately, depending on your opinion of lawyers), it seems like there is no supply shortage of attorneys, so at least that should not be a problem you might encounter.

In Europe, availability of skilled labor is also a substantial obstacle to wide-scale adoption of microgrids. The energy transformation is well underway, with ambitious targets set by the European Union's Fit for 55 plan and the REPowerEU energy security action plan [30]. However, meeting these targets will require a significant acceleration in the deployment of clean energy technologies, which in turn will create significant demand for labor and material [31]. The demand for workers to develop and construct wind and solar installations in the EU is expected to increase by a factor of three to four by 2030 [32]. For example, the use of heat pumps and solar PVs has grown significantly in recent years, requiring an increasing number of installers, plumbers, and electricians [33]. The EU is expected to need up to 18 times more lithium by 2030 and up to 60 times more by 2050 for EV batteries and energy storage alone [34]. Other key materials, such as nickel and cobalt, are expected to become scarce by 2025 [35].

Examples of Supply Chain Issues for the EU

Solar Photovoltaics: The REPowerEU plan calls for the installation of 320 GW of new solar PV by 2025 and 600 GW by 2030. Achieving this goal will require a threefold to fivefold increase in solar PV installation rates [36].

Raw Materials: The EU is seeking to increase its upstream capacity and that of neighboring countries, with new mining projects under development, such as a lithium mining project in the Czech Republic, and a lithium borate project in Serbia [37].

To meet these challenges, companies and stakeholders need to analyze and strengthen supply chains, invest prudently, and take mitigation measures to ensure supply chain resilience. This approach can help Europe seize the significant opportunities of an orderly energy transformation and meet its ambitious energy goals.

Concern #4: Money

While we list this as the last of the concerns in this chapter you must consider when embarking on this path, it is by no means the least of your concerns or the last thing you should consider. This concern pervades every step of this effort. How a budget is planned and managed can make or break a project regardless of how successful you have been at addressing the previous concerns and risks. In many ways, this is probably the most complex of matters to address, especially if there are multiple stakeholders or owners involved in planning and funding the system.

First, you must decide whether to purchase or finance the system. There may also be options to lease it or set up a PPA. The PPA is probably the easiest to finance, since all financial risk related to construction, permitting, regulations, etc. is borne by the installing and owner/operator company(ies), but it would cede all ownership of the underlying asset to whoever financed the system, leaving the planners at risk of disruption should rate terms expire and have to be renegotiated or the system fails and the company that installed and maintains it also fails. In the case of both a PPA or a financed agreement, true autonomy has not been achieved, since the system end users are not in control of their assets, at least not until all debt payments have been made in the financed scenario. So to be truly autonomous from the outset, the system needs to be paid for in cash. And that, in most cases, will amount to a handsome sum.

We will not delve into the intricacies of PPAs for project financing here, since those topics are tangential to the subject of this book and are also well established commercial and banking practices better addressed by the institutions you might engage. Instead, let us assume you want to establish full autonomy. You want total ownership and control of the system and the freedom to decide how you contract for its implementation and operation or, where possible, self-perform. If you are a corporate microgrid installer, opting not to finance would be odd, since few companies wish to dedicate that much cash to projects like this unless the return is expected in the relatively short term. If you happen to be a cash-rich company, and there are not better-yielding opportunities for you to employ that cash, then this might be a good, and relatively easy option for you. Let us consider the more complicated scenarios—those where the microgrid would benefit multiple stakeholders of varying financial means, like a small community for example.

In the community scenario, the funds will come from either individual contributions or from the district/town/neighborhood tax base or other kinds of assessments (e.g., a homeowners association assessment or some kind of membership assessment if it is not from local taxes). The cleanest way to do this that would generate the least angst among stakeholders would be by an assessment that is applied equally to all (e.g., similar to a property tax but based on the average electrical load of an individual's assets instead of the land area they own). That way, there would not be many, if any, who feel more or less entitled to the system's benefits. At the end of the day, the funding path you choose should ultimately be what is best for you to achieve the results you desire. The next problem is the timing and cost estimating.

Once upon a time, in a reality that now seems distant but in fact is not, it was possible to plan significant construction projects of considerable technical complexity, like microgrid installation, with a high degree of cost estimating confidence. Budgets could be set based on cost estimates that could be verified and expected to remain valid for long enough to get the funding, permits, contracts, etc. in place to begin work and then execute the project within that planned budget. That world, which seems like a fading dream, no longer exists. Macroeconomic upsets that result in sudden surges in material or labor cost are now the norm. Inflation in many parts of the world remains relatively high. As mentioned previously, supply chains for the products needed for these projects remain strained. The only thing that remains certain is that uncertainty will be inherent to any plans. For this reason, we strongly suggest you set a base case budget scenario and a worst-case scenario and budget according to the worst-case scenario. Have a wish list of things you will do with any leftover money if macroeconomic trends move in your favor, but the worst thing that can happen is to be ready to begin work and not have enough money to complete it.

European Perspective

In Europe, financing clean energy projects is a complex task involving different mechanisms and institutions. The EU has established several financing mechanisms to support renewable energy projects and to promote the increased use of renewable energy sources in the EU [38].

EU Renewable Energy Financing Mechanism

This mechanism enables EU countries to work more closely together in the deployment and promotion of renewable energy. It allows two or more countries to jointly support new renewable energy projects by providing funding and land.

Connecting Europe Facility (CEF)

The CEF is the EU's financing instrument for the promotion of energy, transport and digital infrastructure. In 2018, the CEF was renewed for 2021–2027 with a budget of €42.3 billion to support investments in EU infrastructure networks for energy (€8.7 billion), transport (€30.6 billion), and digital (€3 billion) [39].

The European Investment Bank (EIB) and the European Fund for Strategic Investments (EFSI)

The EIB helps finance energy projects by providing loans and other financial instruments to companies. The EFSI is a joint initiative of the EIB Group and the Commission to mobilize private investment in projects of strategic importance to the EU, including energy efficiency, renewable energy, power grids, and interconnections [40].

Examples of Microgrids Around the World

This section highlights examples of microgrids being used around the world. It describes what they can do, the benefits they bring and the problems they solve. These examples, which include: community-based projects in Australia and Italy; innovative hybrid systems in India and Tanzania; resilience and energy savings solutions on US military installations; and rural community resilience and emission reducing systems, show how microgrids can make energy systems more reliable, reduce our reliance on traditional power sources, and support local economic development.

There are many other existing and planned microgrids we could present, and more are coming online or breaking ground every day. These are but a small set of examples we chose to cite to demonstrate the diversity of use cases, configurations, and end user types that have been contemplated and implemented.

Example #1 [41]

End User: United States Air Force (USAF)

Location: Hanscom Air Force Base, Massachusetts, United States

Configuration: 4.6 MW natural gas fired combined heat and power (CHP) including cybersecurity, a microgrid control system, demand response, distributed generation, non-tactical fleet electrification.

Purpose: The plant generates enough electricity to power 40–50% of the base during a power outage. Additionally, it provides resilient power to mission-critical facilities, bolstering the base's resiliency.

Construction Cost of CHP: $20.5 million (in 2020)

Business Structure: third-party owned and operated system in an Energy as a Service (EaaS) contract with the USAF.

Example #2 [42]

End User: Medley at Southshore Bay housing development

Location: Hillsborough County, Florida, United States

Configuration: 7.8 kW of rooftop solar and 17.7 kWh of battery storage per home on 37 homes with 240 kWh of battery storage and a pair of reciprocating natural gas generators totaling 350 kW at a "community energy park" for backup power.

Purpose: Each home in the pilot program will have utility-owned rooftop solar panels and a BlockBox, containing battery storage and distributed controls, which is about the same size as an air-conditioning unit. Each home's BlockBox will connect to a neighborhood-level loop system, which shares power with all the homes in the microgrid. Through the project, Tampa Electric (TECO) has tested the neighborhood microgrid's ability to withstand AC distribution system disturbances, integrate renewable energy sources and reduce stress on the utility's transmission and distribution system during times of peak demand. The Pilot has demonstrated its ability to maintain electric service during Tampa Electric's 42-hour AC system outage event during Hurricane Ian. During the outage, the BES maintained service while the surrounding homes experienced an extended outage.

Construction Cost: $2 million (in 2021) in costs authorized by Florida Public Service Commission

Business Structure: TECO owned and operated infrastructure

Example #3 [43]

End User: Alutiiq Tribe of Old Harbor

Location: Old Harbor, Kodiak Island, Alaska, United States

Configuration: This project plans to construct a run-of-the-river hydroelectric facility on Kodiak Island, Alaska capable of generating about 3,470 MWh of energy annually and offset diesel fuel use at the local power plant by 95%, enabling a year-round reliable energy source.

Purpose: Old Harbor is a remote Alutiiq Tribal village on Kodiak Island, Alaska, approximately 50 air miles from the nearest social services in the City of Kodiak. There are nearly 200 residents in Old Harbor and 83% are Native Alaskans. The Tribe intends to have 100% of the proposed financial benefits go directly to Old Harbor residents as water, sewer, and electric bill subsidies, helping to reduce energy burdens of the community.

Construction Cost: TBD, up to $10M cost share from US Department of Energy

Business Structure: Locally owned and operated, financed by a US Department of Energy grant.

Example #4 [44]

End User: City, businesses and residents of the City of Montezuma

Location: Montezuma, Iowa, United States

Configuration: The project aims to deploy a 2.5 MW solar array and 1.5 MWh battery energy storage system (BESS) microgrid with EV chargers and advanced metering to reduce reliance on aging infrastructure and back-up diesel generation.

Purpose: This project plans to implement the first utility-scale microgrid in the state, providing energy resilience and reliability and reducing greenhouse gas emissions. Montezuma is a rural community with 1,460 residents that experience extreme weather events, aging infrastructure, and a 53% reliance on fossil fuels.

Business Structure: $12 million primarily funded by a US Department of Energy grant, with contributions by Iowa State University and local utility Montezuma Municipal Light & Power.

Example #5 [45]

End User: City, businesses, and residents of Borrego Springs, California

Location: Borrego Springs, California, United States

Configuration: The microgrid consists of San Diego Gas & Electric Company's 69 kilovolt to 12 kilovolt air-insulated substation; a 26 MW PV system located in the community; approximately 3 MW of customer-owned rooftop PV systems, two 1.8 MW distributed generation resources; two substation batteries; and three community energy storage batteries to provide electricity to the entire community during the day and, when necessary, dropping to critical loads at night.

Purpose: To develop a system to island the Borrego Springs community and eliminate much of its vulnerability to interruptions of the electric grid. Expanding the existing microgrid facility boundaries to energize all the Borrego Springs customers by adding synchronizing relays, integrating third-party PV systems, hardening of electric overhead structures, and field automation. Developing an advanced controller that could successfully island the entire Borrego Springs community and all its customers. Controlling the microgrid locally and from SDG&E's Distribution Control Center more than 80 miles away.

Business Structure: Owned and operated by the utility San Diego Gas & Electric.

Example #6 [46]

End User: Bad River Band of Lake Superior Tribe of Chippewa Indians reservation's Odanah, Aspen Acres, Franks Field, and Birch Hill communities

Location: Bad River Reservation, Wisconsin, United States

Configuration: 5 MW solar photovoltaic (solar PV) array and an 8 MWh battery energy storage system to connect with existing diesel and natural gas generation

Purpose: Located along the shores of Lake Superior, residents of the Bad River Reservation experience harsh winter weather, extreme storms, and flooding known to cause power outages. The implementation of a hybrid mini grid would increase the Tribe's energy resilience and reliability, while moving them closer to energy sovereignty. This project aims to help the Mashkiiziibii reach its goal of carbon free emissions with 100% renewable electricity generation by 2027 and provide a model for replication in other tribes and rural communities.

Community Microgrid Examples

The following examples are described in a contextual format to highlight the needs addressed by the microgrid, as opposed to the project summaries provided above that focus on the business cases of those projects. These examples show the important effect local energy systems can have on communities all over the world. By using renewable energy sources and advanced technologies, microgrids provide long-lasting and reliable power solutions that are adapted to the specific needs of each area. These projects improve energy security and reduce environmental impact, as well as encouraging community engagement and economic growth. As the global energy landscape continues to change, microgrids will play a crucial role in driving the transition towards a more sustainable and decentralized energy future.

Northern Perth, Australia [47]

In the north of Perth, 119 households worked together on an energy project. They shared a battery that could store 1.1 MWh of energy. This meant that everyone was able to save over AUD 81,000 in five years, and it also meant that the local electricity network did not have to put up with as much demand during the day, with an 85% drop in electricity being taken from the grid at the busiest times.

The shared battery system allowed for efficient energy storage and distribution, making the most of solar power generated locally. The project also included the installation of solar panels on individual homes, which contributed to the overall energy generation. The community was involved in managing and maintaining the battery system, which made people feel like they owned it and worked together, leading to more sustainable energy practices.

Magliano Alpi, Italy [48]

The energy community of Magliano Alpi in the Italian Alps developed tools to predict how much energy would be produced and how much would be needed. This allowed the community to effectively use their solar PV systems. The project meant that the community could cover 35% of their electricity needs, meaning they did not have to rely on the national electricity grid and could wait to pay for expensive infrastructure upgrades. The solar PV systems installed have a combined capacity of 200 kW, which has made a big difference to the community's energy independence. They also used LED lighting and energy-saving appliances to consume even less energy. The project also included educational programs to raise awareness about renewable energy and sustainability among residents.

Denmark – Samsø Island Microgrid [49]

The island of Samsø in Denmark is now using 100% renewable energy, such as wind turbines, solar panels, and biomass energy. The community owns and operates these energy resources, which include 11 onshore wind turbines with a total capacity of 23 MW, solar panels with a capacity of 2 MW, and biomass plants generating 10 MW. The microgrid makes the community energy independent and also makes money for the community by selling excess energy. The Samsø Energy Academy is an important part of this process, as it is a place where people can learn about, train in, and get involved in the community. The academy offers workshops, tours, and seminars to teach residents and visitors about sustainable energy. The island's success is all down to this strong community spirit and cooperative tradition.

Feldheim, Germany

Feldheim is a great example of a community that works together to produce its own energy. This small village of about 130 residents is located in Brandenburg. It has become fully energy self-sufficient through its own microgrid. This has been developed and managed cooperatively by the villagers. The local energy system includes 47 wind turbines (producing 74 MW), a 2.3 MW biogas plant fueled by livestock waste and crops, a 2.25 MW solar farm, and a wood-chip heating plant. Electricity and heat are supplied through local grids that are not connected to the national grid. The community created its own grid infrastructure with support from public and EU funds, allowing residents to access stable, low-cost energy entirely from renewable sources. Feldheim's model not only greatly reduces carbon emissions but also helps the local economy to grow and makes people feel like they have a share of the energy and are proud of it [50].

Redwood Coast Airport Microgrid, California [51]

The Redwood Coast Airport Microgrid (RCAM) in Northern California uses solar power, battery storage and a microgrid controller to provide reliable, renewable energy to the airport and the local community. The solar panels can produce up to 2 MW of electricity, and the battery can store up to 8 MWh of energy. This project is an example of how local communities can use renewable energy and shows how well this kind of project could work in the future. The RCAM project is a joint effort between the Redwood Coast Energy Authority and Humboldt County. As well as making energy systems more resilient, the project supports local economic development by investing in renewable energy infrastructure and creating jobs. The microgrid also makes the airport more efficient and reduces its carbon footprint.

Europe: The Netherlands: Aardehuizen Microgrid [52]

The Aardehuizen microgrid in Olst, Netherlands, is a community-based microgrid with 23 homes. It is designed to be sustainable and self-sufficient. The community uses solar panels with a total capacity of 150 kW, EVs, heat pumps, and batteries to create a decentralised energy network. Each home has a smart meter, and the neighborhood is connected to a single transformer. This means that the community can operate independently from the main grid during outages. The microgrid can meet 90% of the community's energy needs and aims to become fully self-sufficient. The project also includes rainwater harvesting systems and organic gardens, which help the community to be more sustainable in all respects. The community regularly holds workshops and events to share knowledge and experiences related to sustainable living.

Europe: Germany: Mannheim-Wallstadt Microgrid [53]

The Mannheim-Wallstadt microgrid in Germany is an intelligent urban microgrid that combines renewable energy sources with energy storage and smart grid technology. The microgrid has solar panels that can produce 500 kW of electricity, battery storage systems that can hold 1 MWh of electricity, and advanced control systems that manage energy use and balance supply and demand in real time. The aim of the project is to provide a reliable and sustainable energy source for a residential area, thereby reducing the community's carbon footprint and enhancing energy security. The smart grid technology lets us see and adjust how energy is used in real time, making sure we use our resources in the best way. The project also includes electric vehicle charging stations and energy-efficient building designs, which further promote sustainability.

United States: Blue Lake Rancheria Microgrid [54]

The Blue Lake Rancheria Microgrid in California is a leading example of a resilient and sustainable energy system. Located in a region prone to natural disasters, this microgrid integrates a solar array, battery storage, and advanced control systems to ensure continuous power. The microgrid serves tribal government offices, a hotel, casino, and other facilities, and serves as an emergency evacuation site for the American Red Cross. The Blue Lake Rancheria microgrid generates significant annual electricity savings and has been recognized for its contributions to energy resiliency and sustainability.

United States: Alcatraz Island Microgrid [55]

The Alcatraz Island Microgrid in San Francisco Bay has solar panels that can produce 305 kW of electricity, battery storage systems that can hold 1 MWh of electricity, and diesel generators to provide reliable power to the island. This project has made the island much less dependent on diesel fuel, which has reduced emissions and operating costs. The microgrid makes the island more energy secure and helps it to meet its sustainability goals. It also includes educational programs to teach visitors about renewable energy and sustainability. The microgrid's success shows that renewable energy can be used in remote locations.

Hybrid Power Microgrid in Baharbari Village, Bihar, India [56]

Baharbari Village in Bihar, India, had a lot of problems with its electricity supply. The villagers' main job is farming, and they had problems with a power supply that was not regular. This affected their irrigation, the processing of farm products, and other activities. To solve these problems, a hybrid power microgrid was set up to provide reliable and sustainable energy solutions. The first plan was to build a biomass gasification plant and a 100% gas engine generator plant. However, this approach resulted in high generation costs due to high maintenance costs and underutilization. To make the microgrid more reliable and affordable, a hybrid system was created. This new system uses solar panels to generate electricity. The hybrid microgrid system included a biomass gasification plant that used locally available biomass raw materials to generate electricity, solar PV panels that used the solar irradiation potential of the region to supplement the biomass-generated power, and energy storage systems that integrated battery storage to ensure a continuous power supply during periods of low generation. The village benefited from a reliable power supply, which made sure that things like irrigation pumps, agro-processing machines, shops, small businesses and households always had power. It also helped the village's economy to grow by powering businesses that make money. It also helped women to be more productive and improve their lives. It supported social needs by providing electricity for lighting, pumping, and other essential services, improving the quality of life for villagers.

Hybrid Power Microgrid in Tanzania [57]

By the end of 2018, only a quarter of the rural population had access to electricity [2]. To solve this, they have started using hybrid power microgrids. The system uses solar PV panels, biomass, hydro, and diesel generators to make sure there is always enough power. As of today, Tanzania has 209 recognized mini-grids in operation. These mini-grids collectively have a capacity of 231.7 MW, which represents approximately 15% of the nation's total energy capacity of 1,461 MW. The energy storage system included batteries to make sure there was a stable power supply even during times when there was not much generation. The hybrid microgrid has provided several benefits to rural communities in Tanzania. It makes sure that people's homes, small businesses and farms always have electricity. This reliable power supply has helped villagers generate and grow their own businesses. It also supports social needs by providing electricity for lighting, pumping, and other essential services, improving the quality of life for villagers.

Hybrid Power Microgrid in Beijing, China [58]

The capital city of China, Beijing, has had big problems finding ways to provide enough green energy for its large housing estates. To solve these problems, a hybrid power microgrid was set up to provide reliable and sustainable energy. This system uses different types of energy sources. The microgrid could produce about 47 MW, and solar panels were a big part of this. The energy storage system included batteries to make sure there was a stable power supply even during times when there was not much generation.

The hybrid microgrid has provided several benefits to the residential community in Beijing. It makes sure that households, small businesses and community facilities always have electricity. This reliable power supply has helped local businesses to generate income and grow. It also supports social needs by providing electricity for lighting, heating, and other essential services, improving the quality of life for residents. However, setting up the microgrid was not without challenges, such as the fact that the demand for electricity was not consistent and it was expensive to run. The community's electricity use varies a lot during the day and night, which makes it hard to provide a reliable and profitable supply of electricity all year round. To make sure the microgrid could make enough money to pay for itself, it was important to plan carefully and make sure the costs of generating electricity were balanced with the prices of electricity.

Hybrid Power Microgrid in Qatar [59]

Qatar is one of the world's countries with the most fossil fuels. It has been looking at ways to use renewable energy to power its desert farms, which are far away from other buildings. These farms often use diesel generators because they are in remote places, so having a microgrid with renewable energy is a good idea. The Qatar Environment and Energy Research Institute (QEERI) has set up a hybrid power microgrid at one of these desert farms. This system uses solar panels, batteries, and a backup diesel generator to make sure there is always enough power. The microgrid was designed using HOMER Pro software to determine the best combination of energy sources based on the farm's energy needs and the area's solar potential. The microgrid can produce 35 kW of electricity from solar panels on the ground, and it also has an 82 kWh battery storage system. A backup generator that runs on diesel ensures that there is always power, even when there is not much solar power. The hybrid microgrid provides several benefits to the desert farm, including a stable and continuous power supply for irrigation, cooling systems in greenhouses, and small-size reverse osmosis-based water desalination plants. This reliable power supply supports agricultural activities, which generates income and helps the economy to grow. It also supports social needs by providing electricity for essential services, improving the quality of life for farm workers.

Addressing Energy Poverty through Microgrids

Not having enough energy is a big problem around the world. As of 2020, about 733 million people did not have electricity [60]. This problem is especially bad in remote and isolated areas, like the Brazilian Amazon, where about 155,000 households still don't have electricity [61]. These communities often use diesel generators, which are expensive, damage the environment, and can stop working when there is a problem with the supply.

Due to diesel's cost volatility, the environmental and air quality impacts, and supply stability concerns during grid outages and after catastrophe as it is a bad choice for a primary energy source. Not having access to modern energy services is a big reason why people leave rural Amazonian communities and move to larger cities. If people had better electricity access locally, they might want to stay in the area, which would protect the forest and help to stop things like illegal logging and mining. Microgrids are a good alternative to traditional diesel generation. They provide reliable and sustainable energy solutions for small and medium-sized enterprises (SMEs), as well as small communities and towns. In small communities or towns, microgrids are usually set up in people's homes to make sure they have their own electricity supply. This involves putting solar panels on roofs and other places like parking areas, and having energy storage systems in open spaces, buildings, or underground.

This means that people who use this energy become prosumers which might not appeal to everyone. It is very important to understand and deal with the reasons why people might not accept smart grid technologies if we want these projects to be successful. If people are not interested and fully committed, it is much harder for the project to be successful.

In addition to the rules, we need to look at how the technology works. Intelligent algorithms are used to manage system operations, making these grids "smart". Users can get involved by using smart meters, which collect data on how much energy their home uses and when their appliances are on. These meters support the integration of various physical elements within the microgrid, including solar panels, inverters, energy storage, electrical loads, and control systems.

To develop energy applications that can adapt to different situations, such as negotiating tariffs, trading energy, and maintaining grid stability, microgrids use advanced AI algorithms and cooperative distributed models. It is very important that all the different people and organizations involved work together to create a digital framework for smart grids and renewable energy that can last.

Using microgrids to tackle energy poverty not only improves access to reliable and affordable energy, but also supports economic development, enhances quality of life, and promotes environmental sustainability. By using modern technologies and getting the community involved, microgrids can be a key part of improving energy access and helping to make the switch to renewable energy.

Microgrids Configuration

A solar microgrid system in the US, for example, typically consists of solar panels, a battery storage system, and a microinverter. The micro inverter converts the direct current (DC) produced by the solar panels into alternating current (AC) that can be used by a variety of electrical loads. The supply of electricity from a solar microgrid system depends on several factors, including the size and efficiency of the solar panels, the amount of sunlight the panels receive, and the capacity of the battery storage system. The consumption of electricity from a solar microgrid system depends on the electrical loads dependent on the system. The level of self-sufficiency of a solar microgrid system depends on the balance between the system's production and the household's consumption. If the system generates enough electricity to meet the household's needs, its self-sufficiency rate can be 100%. In the US, most microgrids are not 100% self-sufficient and remain interconnected with the grid for the remainder of the supply needed, and to have a backup supply should part of the microgrid fail. A series of useful links to microgrid planning resources are provided in the footnote [62].

Social Acceptance

Experience shows that if most inhabitants in an area do not support a clean energy project, it can be challenging to proceed. In France and in the US, for example, the installation of windmills has faced very vocal opposition from locals due to concerns about the destruction of their view and noise pollution. Therefore, it is essential to introduce early to affected people the concept of them becoming part of an energy community.

Social acceptance regarding microgrids [63] involves the willingness of individuals, communities, and societies to adopt and support the integration of these systems into their daily lives. To achieve social acceptance, it is important to use clear, concise, and objective language that avoids biased or emotional expressions when describing the reason for implementing a microgrid, its purpose and the benefits it will provide. Social acceptance is crucial for the successful implementation of microgrids, renewable power and other smart grid technologies.

In both the US and Europe, social acceptance is influenced by several factors. First, it is important for people to have a clear understanding and awareness of what a microgrid is, how it works, and the benefits it can provide, including energy independence, potential cost savings, and environmental benefits. Second, trust is crucial. The reliability of the microgrid and the operator's ability to provide effective service and support are essential for gaining trust in the technology and the companies that provide it. Government regulations and incentives can also significantly impact social acceptance of microgrids, by making them more affordable and less perceived as a luxury only available to the wealthy. Visual appearance, any noise related to operation, and potential inconveniences during installation (e.g., traffic problems, noise, dust, power supply interruptions) should be considered. Moral values such as privacy or environmental justice can also impact the acceptance of smart grid technologies. It is important to note that social acceptance can vary widely from one individual or community to another and is influenced by a complex interplay of factors. Therefore, strategies to increase social acceptance should be multifaceted and tailored to specific contexts.

Context of Social Acceptability

Smart grids provide real-time information on energy needs. Because microgrids are a form of smart grids that possess their own energy generation, a citizen owner/operator of a microgrid transforms from a pure energy consumer to an energy prosumer. This affords them the options to delay energy consumption, store energy that has been produced or purchased at a discounted price and sell it later for a profit or even donate energy to the community. The concept of energy communities [64] is relevant because as a prosumer, one cannot simply be individualistic as was the case when one was a pure consumer. Additionally, an energy community needs to be agile to address variability in energy production, such as when there is no wind or sun, or when extreme events force the community to operate in a degraded mode, assuming the grid is also unavailable as a backup source of power. In this context, social acceptability becomes essential, requiring behaviors and mutually-binding commitments based on rules, cooperation, and mutual support for the community to operate harmoniously and sustainably.

Figure 3.1 shows how to encourage interaction to create inclusive microgrid communities. The diagram shows how people in a community can work together to share information, learn about energy, and get more people involved. It shows how people can help make microgrid communities more sustainable and happy by thinking about how important this is and how vulnerable some people are. This approach makes sure that everyone in the community is involved in making decisions, which leads to more fair and strong energy solutions.

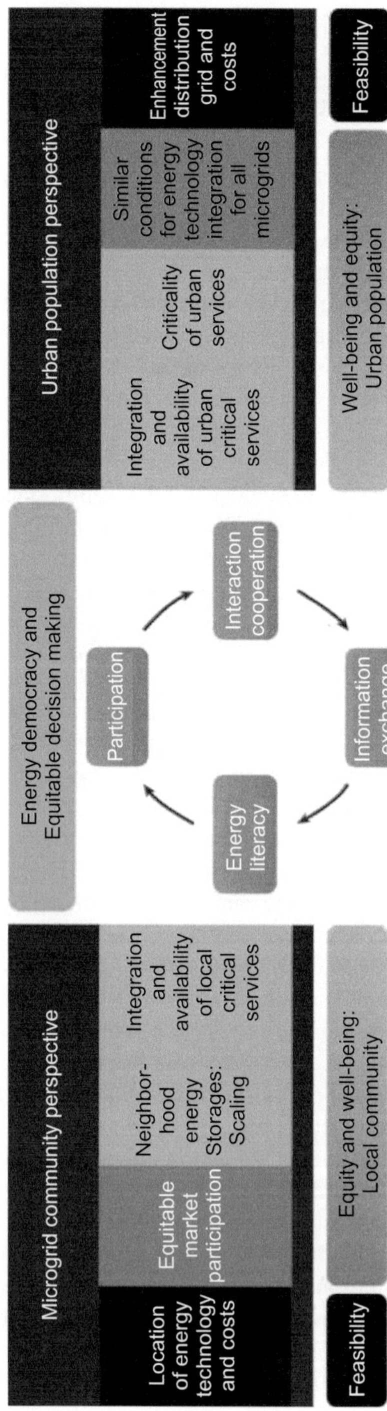

Fig. 3.1 Fostering interaction, a cornerstone for inclusive microgrid communities [65].

Conclusion: Navigating the Path to Energy Independence

As we conclude Chapter 3, it is clear the practical implementation of microgrids requires meticulous planning, technical expertise, and a thorough understanding of regulatory landscapes. This chapter provided a roadmap for engineering your energy freedom, covering the essential aspects of microgrid design, technology selection, project management, and legal considerations.

The chapter began with an exploration of microgrid design and engineering, offering guidance on tailoring a microgrid to meet specific needs and circumstances. It emphasized the importance of a well-thought-out design that integrates various renewable energy sources, energy storage systems, and smart grid components. The selection of appropriate technologies is crucial for ensuring the efficiency, reliability, and sustainability of the microgrid.

Project management emerged as a critical component of successful microgrid implementation. Best practices for managing a microgrid project from inception to completion were discussed, including timelines, budgeting, and stakeholder engagement. Effective project management ensures that the microgrid is built on time, within budget, and meets the desired performance standards.

Navigating the regulatory landscape is another essential aspect covered in this chapter. Obtaining necessary permits and ensuring compliance with local, state, and federal regulations can be complex and time-consuming. However, understanding and adhering to these regulations is vital for the legal and operational success of the microgrid.

Real-world examples of successful microgrid projects were presented, highlighting the challenges faced and solutions implemented. These case studies provided valuable insights into the practical aspects of microgrid deployment, demonstrating how various communities and organizations have achieved energy independence and resilience.

As we move forward, it is important to recognize that the journey towards energy autonomy is an ongoing process. Continuous monitoring, evaluation, and adaptation are necessary to ensure the long-term success and sustainability of the microgrid. By staying informed about the latest technological advancements and regulatory changes, communities and organizations can optimize their microgrid systems and maximize their benefits.

With a comprehensive understanding of the practical steps required to build and operate a microgrid, we are now ready to explore the broader implications of energy autonomy. Chapter 4, titled "The Who and What Now? Empowering Prosumers and Shaping the Future", will delve into the roles of individuals and communities in the energy revolution.

In Chapter 4, we will examine the concept of prosumers and their impact on the energy landscape. We will explore how technological innovations and data management are transforming microgrids and enabling more efficient and secure energy use. The chapter will also discuss modern economic models and sustainability, highlighting the principles of the circular economy and the reduction of carbon footprints.

Furthermore, we will review international regulations on climate change and the role of the Sustainable Development Goals (SDGs) in supporting smart grids. Operational strategies and change management will be addressed, providing tips for successfully transitioning to energy autonomy. Finally, we will look at future opportunities and business models, exploring new ways to encourage energy efficiency, create jobs, and develop innovative business models for microgrids.

By the end of Chapter 4, you will have a deeper understanding of the social, economic, and technological dimensions of the potential microgrids have to transform the energy landscape. You will be equipped with the knowledge and tools to actively participate in and contribute to the transformation towards a sustainable and resilient energy future.

References

1. *See*, e.g., the mission statements of the California PUC, Florida PSC and Pennsylvania PUC: https://www.cpuc.ca.gov/about-cpuc/divisions/news-and-public-information-office/consumer-affairs-branch; https://www.psc.state.fl.us/about; https://www.puc.pa.gov/about-the-puc/#:~:text=The%20mission%20of%20the%20Pennsylvania,economic%20development%3B%20and%20foster%20new.
2. *E.g.*, Florida Bill SB 796 requiring utilities to harden overhead power lines and burning some power lines. https://www.flsenate.gov/Session/Bill/2019/00796/?Tab=BillHistory.
3. https://www.nuscalepower.com/press-releases/2025/nuscale-powers-small-modular-reactor-smr-achieves-standard-design-approval-from-us-nuclear-regulatory-commission-for-77-mwe.
4. https://www.solar.com/learn/solar-panel-cost.
5. https://nenpower.com/blog/how-much-does-a-solar-panel-cost-per-watt-in-europe/?utm_source=chatgpt.com.
6. https://seia.org/research-resources/solar-market-insight-report-2024-year-in-review/. https://www.woodmac.com/reports/power-markets-us-solar-market-insight-2024-year-in-review-150361048/.
7. https://www.spglobal.com/commodity-insights/en/news-research/latest-news/electric-power/011824-rising-costs-lower-revenues-for-european-wind-and-solar-lift-ppa-breakeven-points.
8. Solar Power Portal. (2025). *MCS: BESS and Heat Pumps Drive Record Year for Small-scale Renewables in 2024*. Retrieved June 8, 2025, from https://www.solarpowerportal.co.uk/mcs-bess-and-heat-pumps-drive-record-year-for-small-scale-renewables-in-2024.
   Energy Saving Trust. (2025). *Solar Panels: Costs, Savings, and Benefits Explained*. Retrieved June 8, 2025, from https://energysavingtrust.org.uk/advice/solar-panels/.
9. *See* "Utilities and the Utilities Sector: Pros and Cons for Investors." https://www.investopedia.com/terms/u/utilities_sector.asp.
10. *See* "Backup Power: Public Implications of Private Substitutes for Electric Grid Reliability". https://www2.oberlin.edu/faculty/pbrehm/BrehmJohnstonMilton_Grid_Reliability.pdf, 2024 and National Academies of Sciences, Engineering, and Medicine. 2023. *The Role of Net Metering in the Evolving Electricity System*. Washington, DC: The National Academies Press. https://doi.org/10.17226/26704.
11. https://www.nrel.gov/state-local-tribal/basics-interconnection-standards.html.
12. NC Clean Energy Technology Center. (2025). *Database of State Incentives for Renewables & Efficiency (DSIRE): Summary Tables – Policies & Incentives by State*. Retrieved June 8, 2025, from https://programs.dsireusa.org/system/program/tables.

13. *See* The Role of Net Metering in the Evolving Electricity System.
14. *Id.* at 40.
15. https://www.epa.gov/green-power-markets/renewable-energy-certificates-recs.
16. https://www.dominionenergy.com/virginia/renewable-energy-programs/net-metering.
17. U.S. Energy Information Administration. (2024). *Annual Electric Power Industry Report, Form EIA-861: Net Metering Data.* Retrieved June 8, 2025, from https://www.eia.gov/electricity/data/eia861/.
18. https://codes.iccsafe.org/content/IPMC2021P1.
19. https://www.inkwoodresearch.com/reports/europe-microgrid-market/.
20. https://build-up.ec.europa.eu/en/news-and-events/news/empowering-energy-communities-path-just-energy-transition-europe.
21. https://www.gevernova.com/power-conversion/news/Empowering-European-Grid-Stability-with-Microgrids).
22. *Source:* https://www.microgridknowledge.com/regions/europe.
23. *Source:* https://www.smart-energy.com/regional-news/europe-uk/mapping-europes-microgrids/.
24. *Source:* https://cordis.europa.eu/article/id/87192-microgrids-for-enhanced-energy-supply.
25. https://www.spglobal.com/en/research-insights/featured/special-editorial/look-forward/labor-a-critical-component-of-supply-chains-under-growing-pressure).
26. International Labour Organization (ILO) & International Renewable Energy Agency (IRENA). (2024). *Renewable Energy and Jobs – Annual Review 2024.* Retrieved June 8, 2025, from https://www.ilo.org/resource/news/jobs-renewable-energy-record-highest-annual-growth-rate-reaching-162.
27. https://www.iea.org/reports/energy-efficiency-2020/energy-efficiency-jobs-and-the-recovery.
28. *Source:* https://www.spglobal.com/en/research-insights/special-reports/look-forward/labor-a-critical-component-of-supply-chains-under-growing-pressure.
29. *Source:* https://www.spglobal.com/en/research-insights/special-reports/look-forward/labor-a-critical-component-of-supply-chains-under-growing-pressure.
30. https://commission.europa.eu/strategy-and-policy/priorities-2019-2024/european-green-deal/repowereu-affordable-secure-and-sustainable-energy-europe_en.
31. https://joint-research-centre.ec.europa.eu/jrc-news-and-updates/do-we-have-sufficient-skills-energy-transformation-changing-labour-market-2024-01-16_en).
32. https://joint-research-centre.ec.europa.eu/jrc-news-and-updates/do-we-have-sufficient-skills-energy-transformation-changing-labour-market-2024-01-16_en.
33. https://www.euronews.com/green/2024/04/17/critical-raw-materials-how-the-eu-hopes-to-secure-a-key-element-of-the-energy-transformation-f.
34. https://www.politico.eu/article/5-things-to-know-about-europes-bid-to-stay-in-the-clean-energy-race-net-zero-critical-raw-material/.
35. https://www.mckinsey.com/industries/electric-power-and-natural-gas/our-insights/building-resilient-supply-chains-for-the-european-energy-transformation.
36. https://joint-research-centre.ec.europa.eu/jrc-news-and-updates/do-we-have-sufficient-skills-energy-transformation-changing-labour-market-2024-01-16_en.
37. https://ec.europa.eu/commission/presscorner/api/files/document/print/en/ip_20_1542/IP_20_1542_EN.pdf.
38. https://energy.ec.europa.eu/topics/renewable-energy/financing/eu-renewable-energy-financing-mechanism_en. https://energy.ec.europa.eu/topics/funding-and-financing/eu-funding-possibilities-energy-sector_en.
39. https://energy.ec.europa.eu/topics/funding-and-financing/eu-funding-possibilities-energy-sector_en.
40. https://energy.ec.europa.eu/topics/funding-and-financing/eu-funding-possibilities-energy-sector_en.

41. https://www.hanscom.af.mil/News/Article-Display/Article/3425925/hanscom-leaders-invest-in-energy-resiliency/.
42. https://www.floridapsc.com/pscfiles/library/filings/2023/05183-2023/05183-2023.pdf.
43. https://www.energy.gov/sites/default/files/2024-08/Factsheet_ERA_OldHarbor_8.14.24FINAL.pdf.
44. https://www.news.iastate.edu/news/2024/02/26/microgrid.
45. https://www.energy.ca.gov/sites/default/files/2021-05/CEC-500-2019-013.pdf.
46. https://www.energy.gov/oced/energy-improvements-rural-or-remote-areas-selected-and-awarded-projects?utm_medium=email&utm_source=govdelivery#nwisconsin.
47. https://www.iea.org/commentaries/empowering-people-the-role-of-local-energy-communities-in-clean-energy-transformations.
48. https://www.iea.org/commentaries/empowering-people-the-role-of-local-energy-communities-in-clean-energy-transformations.
49. Colell, A. (2021). Islands: Samsø and Mull. In: *Alternating Current–Social Innovation in Community Energy* (pp. 131–163). Wiesbaden: Springer Fachmedien Wiesbaden.
50. https://nef-feldheim.info/the-energy-self-sufficient-village/?lang=en.
51. https://sonomacleanpower.org/community-partnerships.
52. https://www.weforum.org/agenda/2018/09/these-dutch-microgrid-communities-can-supply-90-of-their-energy-needs/.
53. https://microgrid-symposiums.org/microgrid-examples-and-demonstrations/mvv-mannheim-wallstadt-microgrid/.
54. https://schatzcenter.org/blrmicrogrid/.
55. https://www.energy.gov/eere/articles/welcome-alcatraz-one-largest-microgrids-united-states.
56. https://hybridpowersystems.org/Madeira2022/wp-content/uploads/sites/26/2022/11/3A_3_HYB22_048_presentation_Nandakumar_Avinash_TN.pdf.
57. https://minigrids.org/wp-content/uploads/2020/06/Tanzania-_Case_Study.pdf.
58. https://microgridnews.com/wp-content/uploads/2018/10/he_2018_Beiging_microgrid.pdf.
59. https://microgridnews.com/qatar-desert-farm-is-powered-by-renewable-energy-microgrid/.
60. https://data.undp.org/blog/1-18-billion-around-the-world-in-energy-poverty.
61. https://ourworldindata.org/worlds-energy-problem.
62. (1) Microgrids across the United States - Clean Coalition
(2) New tech can make your house a solar microgrid | Ars Technica
(3) Grid-Connected Micro-Photovoltaics in Germany
(4) Mini-grid for extended family - Victron Energy
(6) Schneider Electric
(7) Enchanted Rock
(8) Today in Energy - U.S. Energy Information Administration (EIA)
(9) Microgrid Program Strategy | Department of Energy
(10) Germany raises feed-in tariffs for solar up to 750 kW
(11) Solar Power Statistics in Germany 2021 - SolarFeeds Magazine
(12) Solar power in Germany – output, business & perspectives
(13) Recent Facts about Photovoltaics in Germany - Fraunhofer ISE
(14) German solar power capacity in 2022 sees fast growth, still well below ...
(15) Micro Grids - Stromnetz Berlin
(16) Designing Microgrids for Efficiency and Resiliency
(17) Installation/User Manual - SoLocal Energy
(18) Microgrid-forming PV microinverter from Enphase
(19) Microgrid Regulatory Policy in the US | Greentech Renewables
(20) New state scorecard finds microgrids are severely under-supported ...
(21) Most states are doing a less than stellar job at deploying microgrids ...

63. Fragniere, E., Sandoz, S., Abdenadher, N., Moussa, M., Serugendo, G.D.M. and Glass, P. (2023, May). Improving the social acceptability of microgrids. In: *2023 IEEE Renewable Energy and Sustainable E-Mobility Conference* (*RESEM*) (pp. 1–5). IEEE.
64. (1) Moral Values as Factors for Social Acceptance of Smart Grid ... MDPI.
    (2) Improving the Social Acceptability of Microgrids. IEEE Xplore.
    (3) Social Acceptance of Microgrids Dedicated to Electric Vehicle Charging ...
    (4) The Research Agenda on Smart Grids: Foresights for Social Acceptance. MDPI.
    (5) https://doi.org/10.3390/su10082703.
    (6) https://ieeexplore.ieee.org/servlet/opac?punumber=10235975.
65. Ottenburger, S.S., Cox, R., Chowdhury, B.H., Trybushnyi, D., Omar, E.A., Kaloti, S.A., ... and Raskob, W. (2024). Sustainable urban transformations based on integrated microgrid designs. *Nature Sustainability*, 7(8), 1067–1079. https://www.nature.com/articles/s41893-024-01395-7/figures/1.

4 Who Leads the Energy Evolution and What's Next?

This chapter explores the human and institutional drivers of the microgrid revolution. It asks who is best positioned to lead the transition to energy autonomy and what lies ahead. The chapter delves into the diverse coalition of stakeholders shaping this evolution, including policymakers crafting incentives and regulations, businesses seeking resilience and price stability, community organizations advocating for equity, and individual citizens demanding cleaner, more reliable energy. The authors emphasize the growing importance of "energy democracy," in which users become active participants—or "prosumers"—in decentralized systems. Drawing on global examples, the chapter examines how grassroots innovation often precedes national policy and how marginalized communities are pioneering microgrid adoption to address historic energy injustice. The chapter highlights the role of education and public awareness in enabling energy-literate populations to make informed decisions about autonomy and resilience. Technological trends, such as artificial intelligence, blockchain-enabled peer-to-peer trading, and next-generation storage, are examined as forces that will further decentralize control and optimize grid efficiency. The chapter addresses anticipated barriers, such as regulatory inertia, upfront costs, and supply chain vulnerabilities, and offers strategic suggestions to overcome them. The chapter also considers what energy systems might look like by 2050, when smart cities, vehicle-to-grid technologies, and distributed, AI-controlled grids may become the norm. Ultimately, the chapter argues that leadership in the energy transition will not come from a single sector, but rather from collective, multi-level engagement across society.

This chapter elaborates on the concept of the energy prosumer. This helps us understand the role of individuals in the changing energy landscape. The chapter then looks at new technology and data management. It explains how new technology is changing microgrids and how technology helps us use energy better and keep it safe.

We also consider modern economic models and sustainability, examining the principles of the circular economy, and how to reduce carbon footprints through the Four Rs (Rethink, Reuse, Reduce, Recycle). This chapter reviews new technologies that can help the economy and the environment.

Then we look at international rules on climate change and the role of the Sustainable Development Goals for smart grids. We describe wider policies that support the switch to sustainable energy. We also look at how to operate and manage microgrids once they are implemented. This addresses the level of self-sufficiency

DOI: 10.1201/9781032689845-4

of the system, effective change management, and strategies for managing a smart grid like an enterprise. This section gives tips for successfully moving towards energy independence.

The chapter concludes with a look at future opportunities and business models. It looks at new ways to encourage people to use less energy, new jobs for young people and new business models for microgrids. It also discusses the notion of Reference Energy System (RES) to provide a holistic view of the wider energy issue and finally how such a paradigm shift will drive a global social change even more than technological change.

Prosumer Power: Orchestrating the Microgrid Revolution

Individual prosumers and communities of prosumers can help change the way energy is produced and managed. The increasing proliferation of prosumers in our neighborhoods, communities, and cities will make the energy system more decentralized and resilient. This change from passive to active participation by individual people and non-utility businesses in the energy ecosystem is important for increased implementation of microgrids and, thereby, greater energy autonomy in society overall.

Introduction to Prosumer Power

Now that your microgrid is up and running and your energy autonomy is realized, you have become a prosumer. This has important implications for us as individuals, as well as for the communities involved. Community responsibility must be considered. Who is driving this decision and how is it organized?

Managing a microgrid, to include production, storage, operation, and consumption, requires a whole series of necessary monitoring, analysis, diagnosis, and decision-making processes. What will be the governance structure for dealing with microgrid issues? How will pecuniary operational losses or gains be managed? What are the benefits, burdens, and limitations imposed on those involved? All microgrids are smart by their nature, which means that algorithms will largely automate these operational processes. However, for the microgrid to function properly, human involvement is necessary. For example, a company could offer a turnkey smart grid "service" system with real-time monitoring that is done remotely with occasional on-site intervention for repairs and maintenance. Alternatively, a community could buy its own smart grid equipment and software, as well as solar panels, inverters, and batteries, and manage everything itself. Between these two extremes, multiple configurations exist.

Local authorities, utility companies, and community stakeholders usually work together to deal with microgrid issues. Local authorities set rules to make sure microgrids are safe and work well. Utility companies give technical support and help connect microgrids to the main grid. In New York, the Reforming the Energy Vision (REV) [1] initiative supports the development of community microgrids with financial incentives. Community stakeholders help make decisions about the microgrid. They make sure it meets local needs and priorities. This helps everyone get what they need and makes sure the microgrid works well and is sustainable.

Money losses or gains in microgrids are managed through different financial models. For example, incentives can be used to encourage the microgrid to be well maintained. In California, the Self-Generation Incentive Program (SGIP) offers financial incentives for installing distributed energy systems, including microgrids [2]. Cost-sharing agreements help distribute financial risks and benefits. Selling extra energy back to the main grid or nearby communities can also pay for it. For example, the Brooklyn Microgrid project lets people sell extra solar energy to their neighbors. These financial strategies are important for keeping microgrids economically viable and providing long-term benefits to communities [3].

The emergence of prosumers is driving the need for new business models. The traditional business-to-consumer (B2C) model, in which utilities sell electricity to consumers, is undergoing a period of evolution. We are now witnessing the emergence of consumer-to-business (C2B) models, where prosumers sell surplus energy back to the grid or to other consumers. Furthermore, business-to-business (B2B) [4] models may entail microgrid operators trading energy with one another, while consumer-to-consumer (C2C) models could involve neighbors directly trading energy.

For instance, a homeowner with solar panels may sell surplus electricity to a nearby business (C2B), or two neighbors may trade energy based on their respective requirements (C2C). These new models create a more dynamic and decentralized energy market, fostering innovation and efficiency.

The notion of demand management must also be considered here. A microgrid community must think about how much of its own energy consumption to conserve. Establishing a virtuous circle should be the desired end state, where a community takes steps to reduce energy consumption while establishing a new source of clean energy and installing the modern energy management infrastructure necessary for a microgrid.

In this chapter, we also discuss the concept of electricity pricing, as it is a critical component of microgrid generation and consumption controls and key to these new business models. We also identify for readers the training needs and actions they will need to take based on their role and involvement in the chosen microgrid configuration and depending on the level of self-regulation chosen.

At the micro level, local microgrids are small-scale power grids that can operate independently or with the main grid. Local communities or small businesses manage these microgrids and would primarily use renewable energy sources like solar panels and wind turbines paired with battery energy storage. This lets communities make, store, and use their own electricity, reducing reliance on the national grid and making those prosumers more energy efficient.

At the meso level, we look at coordinating multiple microgrids. Microgrids in a region are connected and managed together to balance supply and demand, much like the larger interconnected regional utility grids currently operate at a much larger scale. If one microgrid has extra energy, it can share it with another microgrid that needs it. This helps keep the power supply stable and makes better use of renewable energy. A regional energy co-op could manage several microgrids, ensuring energy flows smoothly between them.

At the macro level, these regional systems are integrated with the existing utility-managed power grids. This grid (or grid of grids) must be able to handle different types of energy from different regions. This requires new technology and rules to make sure everything works well. For example, an overarching energy management system might use AI to predict energy needs and make sure electricity from renewable sources is distributed evenly across the grid-of-grid's territory. This makes sure the whole energy system is reliable and efficient.

- **Dynamic:** Electricity consumption is charged dynamically. Prices are adapted to supply and demand over the current period.
- **Two Fixed Rates:** Electricity consumption is charged on the basis of two fixed rates. One rate for peak hours, between 7 a.m. and 9 p.m., and another, cheaper rate for off-peak hours, between 9 p.m. and 7 a.m.

Community Type

- **Cohesive:** Residents have a strong sense of group cohesion. They hold regular meetings to discuss electricity-related issues and are involved in managing the building's infrastructure. They are also able to compare each others' electricity consumption.
- **Independent:** Residents value their independence and are happy to live on their own. They do not meet regularly to discuss electricity-related issues and are not involved in managing the building's infrastructure. Information on electricity consumption is not shared between residents.

What is Your Persona as a New Prosumer

In marketing, a "persona" is a made-up research-based representation of a group of users. It is used to help design products and services that meet the needs of different people. When we talk about the energy transition, we use this idea to look at the different types of people who are joining energy communities as prosumers. Prosumers are people who produce as well as consume energy.

These personas are based on real data about demographics, behaviors and attitudes. In our case, these are represented by what we call prosumer configurations. Each configuration is based on a particular combination of technology choices, lifestyle values, and energy preferences. These preferences may include whether someone lives in a building with battery storage, prefers fixed or dynamic electricity pricing, or values strong community governance. This shows that there is not just one way to be self-sufficient in energy. It also shows that there are many ways people can get involved in the energy system.

Some personas may prioritize sustainability and environmental conservation, actively seeking out energy-efficient buildings with renewable energy features.

Others may be less concerned about sustainability and prioritize other factors such as affordability or convenience.

Pricing sensitivity is another factor that may influence the choice of an apartment building. Different personas may have varying levels of sensitivity to pricing and cost considerations when selecting a residence. Some may be willing to pay a premium for access to renewable energy and smart grid technologies, while others may prioritize cost savings and seek out more affordable options.

Technological savvy is another factor that may influence the choice of an apartment building. Personas may differ in their familiarity and comfort level with smart grid technologies and digital innovations. Some personas may be early adopters and readily embrace new technologies, while others may be more cautious or hesitant about incorporating technology into their daily lives.

By defining personas, researchers can gain deeper insights into the preferences, motivations, and decision-making processes of individuals within the target audience. This information can be used to inform the development of more effective marketing strategies.

To help you understand prosumer personas better, we will look at two example configurations from the authors' own research. These are shown in Figure 4.1, which compares Building A and Building B—two apartment buildings that take different approaches to energy, technology, and community. The cards show the differences in things like grid connection, electricity bills, information sharing, managing quotas, and how often meetings are held. They show how people can have very different preferences in the same energy system, depending on what is important to them.

This first comparison sets the stage for a more detailed framework of six persona-based configurations developed from the same research project. These six types, shown later in Figure 4.2, represent a range of prosumer profiles [6], each defined by a unique combination of attributes.

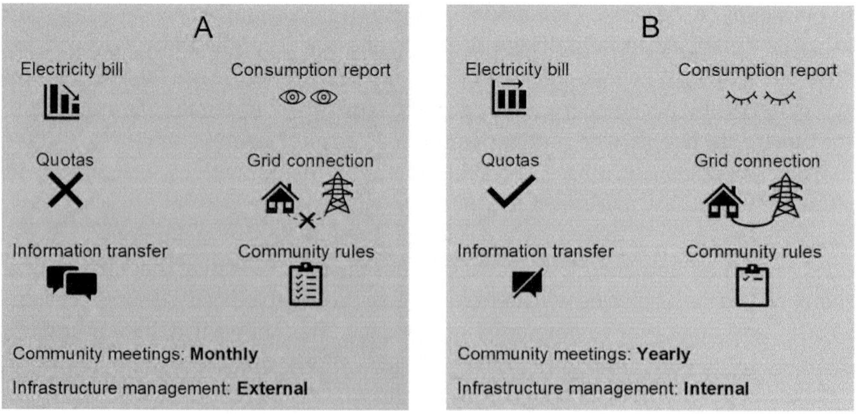

Fig. 4.1 Two sample prosumer configuration cards [5].

Feature	Building A	Building B	Building C	Building D	Building E	Building F
Grid connection	Not connected	Connected	Connected	Not connected	Connected	Connected
Electricity mix	Only renewable	Not only renewable	Only renewable	Only renewable	Not only renewable	Only renewable
Quota	With quota	Without quota	With quota	Without quota	With quota	Without quota
Invoicing	Dynamic	Two fixed rates	Dynamic	Dynamic	Two fixed rates	Two fixed rates
Community type	Cohesive	Independent	Independent	Independent	Cohesive	Cohesive

Fig. 4.2 Prosumers configuration for 6 different buildings.

Before ranking, it is important to understand the attributes that define each persona and how they influence their energy choices and preferences.

Here is a didactic explanation of each attribute, to give the uninitiated a better grasp of the issues in this section:

Network Connection

- **Connected:** The building is connected to the regional electricity grid.
- **Not Connected:** The building is not connected to the grid and is self-sufficient in energy.

This choice reflects the level of energy independence desired by the prosumer.

Electric Mix

- **100% Renewable:** The building's electricity mix is made up entirely of renewable energy, mainly solar PV.
- **Mixed:** The building's electricity mix combines renewable and non-renewable energies.

This attribute indicates the importance placed on clean and sustainable energy sources.

Storage System

- **Battery:** The building is equipped with a battery energy storage system.
- **No Battery:** The building has no energy storage system.

The presence of a battery makes it possible to store energy produced in excess and use it when production is low, thus increasing energy autonomy.

Energy Management System

- **Intelligent:** The building is equipped with an intelligent system that optimizes energy production, consumption, and storage in real time.
- **Basic:** The building has a basic energy management system.

An intelligent system enables finer-grained energy management, promoting energy efficiency and cost reduction.

Electricity Pricing:

- **Fixed:** The price of electricity is fixed, regardless of the time of day.
- **Variable:** The price of electricity varies according to supply and demand.
- **Two Fixed Tariffs:** Two fixed tariffs are applied: A tariff for peak hours and a cheaper tariff for off-peak hours.

This criterion determines the financial impact of energy consumption and can encourage people to change their habits to take advantage of the most advantageous tariffs.

Type of Community:

- **Cohesive:** Residents have a strong sense of cohesion. They meet regularly to discuss electricity-related issues and are involved in managing the building's infrastructure. They can also compare their respective electricity consumption.
- **Independent:** Residents value their independence. They do not meet regularly and are not involved in the management of the building infrastructure. Information on electricity consumption is not shared between residents.

This last attribute reflects the desired level of involvement and collaboration within the energy community.

By analyzing these different attributes, you can classify the six prosumer profiles according to your own priorities. For example, if you place a high value on sustainability and energy autonomy, you might favor buildings with a 100% renewable electricity mix, battery storage, and an intelligent management system.

The aim of this exercise is to demonstrate that there is no single solution for integrating an energy community. Each individual must choose the configuration that best suits his or her needs, values, and lifestyle.

Situation: You're evaluating 6 apartment buildings (A and B), each offering different approaches to energy-related participation and governance.

After reviewing these configurations, you'll be asked to indicate which building you would prefer to live in and to rank the options according to your personal preferences and lifestyle priorities.

Grid Connection

- **Connected:** The building is connected to the regional electricity grid.
- **Not Connected:** The building is not connected to the regional electricity grid. It is self-sufficient.

Electricity Mix

- **Only Renewable:** The building's electricity mix is entirely composed of renewable energies, mainly solar power via PV panels.
- **Not Only Renewable:** The building's electricity mix is not solely composed of renewable energies, but also of other energy sources.

Quota

- **With Quota:** Each household has a quota (quantity of electricity available) that must be respected. The quota is set according to the number of people living in the household and the surface area of the apartment.
- **Without Quota:** No quota (amount of electricity available) is imposed on households.

Invoicing

You can now rank the buildings (see Figure 4.2) in your order of preference, starting with the one you would prefer to live in.

These configuration cards are part of a bigger system called card-based systems, which are often used in conjoint analysis. This helps researchers understand how people choose between different features when choosing an apartment building—in this case, one with different energy setups. By ranking their preferences, participants reveal what matters most to them: sustainability, autonomy, pricing, or community. The most important thing to understand is that there is not one solution that works for everyone. Instead, it shows how people have different expectations and helps the people who make the rules, the designers, and the energy companies to create solutions that meet the real needs of future prosumers.

Technological Innovations and Data Management

Good data management and new technologies help microgrids work better. This section looks at the latest renewable energy, smart grid and data analytics innovations, and how they can improve efficiency, reliability, and sustainability. These technologies help prosumers manage their energy use, which is in line with the chapter's theme of empowering people in the energy transition.

Electricity is the Most Perishable Energy

In the future, electricity will be the most consumed form of energy because it is perfect for digitalization and sustainable development. Electricity is also the most perishable form of energy in that storing it requires complex systems to effectively store and later discharge it vs. the simple tanks we use to store gasoline, propane, diesel, etc. We need to find more ways to store it efficiently and commoditize those new technologies. For example, experiments are currently being done to evaluate the prospects for using EVs as backup batteries for houses, businesses, and possibly for entire energy communities. Also, other battery and storage technologies are being tested, and in some cases scaled up to commercial production, to offset reliance on the most widely used, currently commoditized battery type—the lithium-ion battery.

For example, imagine your EV is fully charged on a Sunday and you don't need to use it. It is a cloudy day, and your smart meter indicates that your household needs extra electricity. In this case, your smart home can use the stored electricity from your EV to meet the household's energy needs. Alternatively, you could buy electricity from the grid, but that could be quite expensive if it is a peak demand time of day and you have variable power pricing. In this instance, it would benefit you monetarily to tap the reserves in your EV, and that would also benefit the grid by reducing peak load and the amount of additional generation (probably from burning coal or natural gas in peaker plants) required by your utility.

Zero-energy houses [6] will also be important in the future. Our homes and cities will eventually become "smarter" than they are now. It sounds like science fiction, but you are already part of the smart revolution with your phone, car, and watch. Your phone's GPS finds you a restaurant, makes a reservation, and orders a taxi. As the appliances and lighting and power systems in your house gradually modernize, your home will similarly facilitate your life and can work with other homes in the microgrid. Houses use a lot of electricity, so research is underway to build homes that produce as much energy as they consume. While zero energy consumption is a practical impossibility, the aim is to create homes that are energy neutral, meaning that they produce enough energy to offset their consumption. The smart grid can predict how much energy we use. New materials, smart meters, advanced controls, storage systems, solar panels, and electric heat pumps will let the house store, produce, and exchange energy according to our needs. The house will help us make smarter energy choices. For example, certain routine processes (EV charging and dishwasher) can be automatically started for when production is at its peak (e.g., midday on a sunny day) or at night when other consumption processes that cannot be delayed (e.g., refrigeration and laundry washing/drying) are likely inactive and the system predicts enough storage remains to power those processes before energy production begins again.

The Smart Logic of Micro grids and AI

We've already explained what a smart grid is. Here, we'll explain how AI is involved in microgrids. This will help us understand the digital twin (virtual microgrid) and swarm logic dissemination sections. Microgrids use advanced tech like AI and ICTs to make energy production, distribution, and consumption more efficient. They use sensors, automation, and control systems to manage energy flows and ensure a reliable supply of electricity. AI helps microgrids work smarter by analyzing data from many sources to predict when energy is needed and adjust how energy storage and other resources are optimally used. This ensures energy is available when needed and stored efficiently when demand is low. AI can also identify problems before they lead to system failure, reducing downtime and maintenance costs.

Microgrids powered by AI offer users several benefits. These systems use real-time data to respond quickly to changes in energy supply and demand, automatically adjusting to maintain balance and prevent blackouts. This is especially helpful in areas that are prone to natural disasters or grid instability. This keeps critical services and infrastructure running during emergencies, making communities more resilient. Microgrids also let users take part in the energy system. User-friendly interfaces let consumers monitor their energy usage, make informed decisions, and participate in demand response programs. These programs encourage users to use less energy during peak periods. This helps balance the grid and reduce costs. AI makes these programs better by giving users personalized advice and responding to the grid automatically. AI can match supply with demand in real time, ensuring efficient energy distribution and maximizing economic benefits for users.

In the future, AI and other new technologies will help microgrids improve a lot. As renewable energy gets cheaper and more people want clean, reliable energy, microgrids will become more common. They will help create a more sustainable energy system that reduces carbon emissions and increases energy security. Virtual utilities and digital twins will also help us make microgrids more efficient and reliable. AI will be key to these innovations, helping to manage complex energy systems in a smart way.

As AI and other innovations develop, microgrids will become a key part of the future energy landscape, providing sustainable energy solutions for communities around the world. Many factors (e.g,. cost reduction, financial incentives, regulatory changes) will affect how these new self-sufficiency technologies spread in the coming years. The government and the level of government involvement in Europe and the USA will be important.

Emergence of Virtual Utilities (Digital Twins)

A digital twin [7] is a virtual representation of a physical object or system, such as a microgrid. Digital twins are used to simulate the behavior of real systems and optimize their performance. For micro smart grids, digital twins can be used to model and analyze the power system, including power distribution, the behavior of individual components (e.g., prosumers and stakeholders), and their interaction with the grid and other power sources. This can optimize the operation of the microgrid, including load balancing, demand response, and renewable energy integration. More and more, this digital twin will enable to first simulate microgrid management and also to manage them in real time. It has to do with the complexity of algorithms and the huge amount of data involved. Digital twins enable engineers to test new smart grid configurations without assuming any risks.

Here are some examples of how digital twins change the way we manage smart grids:

1. **Digital Twins:** They help us test how microgrids will behave in different situations. Engineers can model different grid configurations, energy sources, and demand patterns to optimize performance. A digital twin can show how adding more renewable energy or changing energy storage affects grid stability and reliability.
2. **Real-time Management:** Digital twin technology lets operators manage microgrids in real time. By using data from sensors and monitoring devices, operators can make informed decisions to optimize grid operation and respond to changing conditions. Digital twins can predict changes in demand and adjust energy generation and distribution to keep the grid stable and meet consumer needs.
3. **Managing Microgrids:** This involves complex algorithms for load forecasting, energy optimization, and fault detection. Digital twins let you test algorithms in a simulated grid before using them in the real grid. Engineers can test algorithms in a simulated environment to ensure they work well in the real world.

4. **Digital Twins:** They let engineers test new smart grid configurations without risking the actual grid. By simulating different scenarios, including extreme weather, equipment failures, and cyber attacks, engineers can assess the grid's resilience and develop strategies to address potential risks. This proactive approach helps enhance grid reliability and minimize the downtime, ultimately benefiting consumers and stakeholders.

For instance, a city wants to upgrade its microgrid to accommodate a growing population and increased demand for renewable energy. Engineers can use a digital twin to model the current grid infrastructure, including solar panels, wind turbines, battery storage systems, and consumer demand profiles. The digital twin simulates future scenarios, such as population growth, changes in energy consumption patterns, and fluctuations in renewable energy generation. Engineers can then explore different grid configurations, such as placing renewable energy assets in the best places, implementing demand response programs, and upgrading grid infrastructure to support increased capacity.

Through simulation and optimization, engineers identify the most cost-effective and sustainable solution to upgrade the microgrid while ensuring reliability and resilience. Once the optimal configuration is determined, engineers can deploy the changes to the actual grid with confidence. Virtual utilities powered by digital twins help manage microgrids. They can be used to simulate, optimize, manage in real time, implement complex algorithms, and test without risk. This technology [8] helps create sustainable and resilient energy systems, which will lead to a greener and more efficient future.

For example, a city planning to expand its microgrid to accommodate a growing population and increased demand for renewable energy can use a digital twin to model the current grid infrastructure. This includes solar panels, wind turbines, battery storage, and consumer demand profiles. The digital twin simulates future scenarios, such as population growth, changes in energy consumption patterns and fluctuations in renewable energy generation. Engineers can then explore different grid configurations, such as optimal placement of renewable energy resources, implementation of demand response programmes, and upgrades to the grid infrastructure to support increased capacity.

Through simulation and optimization, engineers can identify the most cost-effective and sustainable solution for expanding the microgrid while ensuring reliability and resilience. Once the optimal configuration is determined, engineers can confidently deploy the changes to the real grid.

Virtual utilities powered by digital twins help manage microgrids by enabling simulation, optimization, real-time management, implementation of complex algorithms, and risk-free testing. This technology helps create sustainable and resilient energy systems, leading to a greener and more efficient future.

"Swarm Logic" for Microgrid Dissemination

In "Star Trek", the Borg are a collective of interconnected units that function as a hive mind. How a grid of microgrids could work is very similar, where multiple energy resources and consumers work together in nearly automatic reaction to changes in the operating environment to make the grid work better. Just as the Borg adapt rapidly to new and changing environments and threats, swarm logic enables microgrids to adjust to changes in energy supply and demand, ensuring stability and efficiency.

The development of national grids followed a centralized model, dominated by major utilities. Microgrids evolve differently, like a swarm. This evolution makes microgrid systems easy to access, which makes them more resilient and allows for controlled expansion. Microgrid swarm logic is different from the centralized model of national grids, which are dominated by major utilities. Each new microgrid addition strengthens the overall energy ecosystem.

Joining a microgrid involves several steps. The new house must have the right wiring and smart meters to fit in with the existing microgrid. The new house must be connected to the microgrid so that its energy production and consumption match the microgrid. This often requires coordination with local energy providers and possibly approval to ensure safety and compliance. This also requires the microgrid to be incorporated as a business entity so it may collectively broker the use and sale of power within and outside the microgrid system. A non-profit energy cooperative business model may be well suited for this type of entity, where excess revenues are reinvested in the microgrid operation or distributed as dividends to microgrid participants. Any financial shortfalls or unplanned capital expenditure would be shared among the microgrid participants, reducing the overall financial burden considerably at the individual participant level.

Once connected, the new house can contribute to and draw from the microgrid. This approach allows for controlled expansion, making microgrid systems more accessible and adaptable. Smart meters help microgrids balance energy production and demand in real time. They also help people use energy wisely by showing how much energy they use and by offering different prices. This same logic can be used in smart grids to automatically find and fix problems, rerouting power to keep the system stable.

Smart meters balance energy production and demand in real time. They also help people use energy wisely by showing how much energy they use and by using pricing models that change with the market. Automation helps the grid respond quickly to changes in energy supply and demand. This also applies to self-healing in smart grids. Self-healing means the grid can fix itself without human help. This means more than just rerouting power. It also means finding and fixing problems like equipment failures or line faults. In this context, smart grids and smart cities work together. Smart cities depend on smart grids, which use city data to make better decisions and use resources more efficiently. This helps optimize energy distribution, enhance reliability, and improve grid resilience.

Also, working together is important for keeping systems stable and reliable. This is because the benefit to society is more important than the benefit to individuals. In short, microgrids moving towards a swarm-like model, using smart technology and working together, could help accelerate the dissemination of smart grids.

Smart Grid and Data Protection

The movie "Live Free or Die Hard" shows how a cyberattack can cause a blackout. This was in 2007, but today it is not science fiction. As we rely more on smart grids and microgrids, our energy systems will become more vulnerable to cyberattacks. Microgrids need to be protected because they use technology that makes them vulnerable to cyber attacks. In recent years, many well-known companies in this field have faced problems because they did not pay enough attention to cybersecurity. They focused on technology and compliance, rather than the bigger picture. After a cyberattack, companies use crisis management techniques like Business Continuity Planning (BCP) to get through the chaos and get back to normal. But while companies protect their systems and data, microgrids face even more challenges. Microgrids operate in an interconnected network that relies on open data sharing. This makes microgrids vulnerable because their data is freely accessible worldwide. In the context of modern AI, which relies on lots of data, this open data environment presents both opportunities and challenges [9].

AI can use global data to make microgrids more efficient and resilient. But the data is more vulnerable to being used by bad people to disrupt the energy supply or compromise the system. Microgrids will be more vulnerable than companies because they rely on open data. This means that data from microgrids may be shared freely around the world. Also, microgrids are connected, so users' privacy and security can be at risk. In a world where personal information is easily accessible, people may be vulnerable to exploitation. The interconnectedness of microgrids can lead to privacy breaches and security threats. As we change how we use energy and use AI, we need to find a way to use open data to help us and keep people and systems safe. This means having good cybersecurity, strong data protection, and a plan to deal with new threats in the fast-changing world of microgrids. Modern AI needs to learn from data gathered around the world to regulate microgrids efficiently while these systems must be protected from cyberattacks [10].

Good data management and new technologies help microgrids work better. This section looks at the latest innovations in renewable energy, smart grids and data analytics, and how they can improve efficiency, reliability, and sustainability. These technologies help prosumers manage their energy consumption, which is in line with the chapter's theme of empowering people in the energy transition.

The integration of AI and machine learning into microgrid management is paving the way for increased optimization of operations and better forecasting of energy production and consumption.

Advanced Applications of AI and Machine Learning

- **Renewable Energy Production Forecasting:** AI can analyze historical and real-time weather data to predict solar and wind energy production, optimizing energy storage management and ensuring a stable supply of electricity.
- **Demand Management:** Machine learning can identify users' consumption patterns and automatically adjust appliances and systems to reduce energy consumption during peak periods.

- **Predictive Maintenance:** AI can monitor the condition of microgrid equipment and predict potential faults, enabling maintenance planning, and minimizing service interruptions.

Cybersecurity in Microgrids

The security of microgrid systems is essential to ensure the reliability and resilience of the energy supply. The decentralized nature of microgrids, with multiple entry and exit points, makes them potentially vulnerable to cyberattacks.

Protection Measures for Microgrid Systems

- **Strong Authentication and Access Control:** Limit access to microgrid control systems to authorized individuals and use strong passwords and multi factor authentication.
- **Data Encryption:** Protect sensitive data in transit and at rest using robust encryption protocols.
- **Monitoring and Intrusion Detection:** Implement monitoring systems to detect suspicious activity and intrusion attempts.
- **Regular Software Updates and Security Patches:** Keep micro-network systems and software up to date with the latest security patches to correct known vulnerabilities.
- **Cybersecurity Awareness and Training:** Train microgrid operators and users in cybersecurity best practices.

Nuclear Energy is Making a Comeback

Nuclear energy is back in vogue, and it has big implications. Nuclear power can meet some of the huge energy needs of AI, but it's a lot easier and faster to build a datacenter than it is to build a nuclear reactor, even the current SMRs being planned. There are also many challenges to scaling up energy production through nuclear integration. Microgrids must now accommodate the complexities of nuclear power. This means making changes to technology, in particular regarding the handling of nuclear waste and getting past public opposition and rules that have made it hard to start nuclear energy projects. This latter statement is very true for Europe.

Nuclear power is also expensive. SMRs are promising but currently require a big upfront investment because they are not yet in commercial production. Google's agreement with Kairos Power to use nuclear energy for its AI needs shows how tech leaders are leading this charge. However, such solutions may be out of reach for smaller operators due to the high costs. Regulations and social issues may make it difficult to deploy new nuclear energy assets, especially in Europe. The licensing, safety, and waste management rules can be hard to navigate. People also often oppose nuclear energy because of safety and environmental concerns. This opposition could slow down or stop the use of nuclear-powered microgrids in some parts of the world, particularly in Europe, even though they could help meet the growing energy needs of AI [11].

Microgrids are designed to improve energy resilience, establish some degree of energy autonomy, reduce costs over time to end users and, in many cases, integrate renewable power to try to mitigate the impacts of climate change. A new factor is changing this landscape: the rise in electricity demand driven by AI. AI systems currently use much more energy than traditional algorithm-based computing systems. It is estimated that by 2028, AI will account for nearly 19% of the total power demand from data centers [12]. This is putting a lot of pressure on existing energy infrastructure and making nuclear energy look like an attractive, carbon-neutral solution again. Nuclear energy is being actively planned as a generation source for datacenter microgrids.

Microgrids have been generally seen as a way to switch to cleaner energy, mainly by using renewable power. But AI is using a lot more energy than people thought, and the pace of deployment and integration of renewable sources is not sufficient to keep up with the increasing power demand. Tech giants like Google, Microsoft, and Amazon are therefore looking at nuclear power and signing agreements to deploy small modular reactors (SMRs) and even recommission mothballed traditional reactors (e.g., Three-Mile Island in New York State and the Palisades reactor in Michigan). This would have been unthinkable just a few years ago. These companies are specifically looking to establish nuclear-powered microgrids to support the datacenters driving their growing AI operations [13].

Modern Economic Models and Sustainability

This section looks at how microgrids can use sustainable energy in a way that benefits the economy by leveraging the circular economy, reducing carbon footprints, and using renewable energy. Modern economic models help communities make money while minimizing environmental impacts.

Modern Concepts of Economic Models

In our exploration of modern economic models related to microgrids, we have discovered many new ideas and reasons for change. As we embrace new ideas like living in energy communities, becoming prosumers, and trying to be energy independent, we see that the current economic paradigm will eventually become obsolete. We must base our goals on new and evolved economic models that promote sustainability, resilience, and equity.

These models are based on the circular economy, which is about short cycles, efficient waste management, reusability, and recycling. This way of thinking values resources and reduces waste. A microgrid lets houses make their own electricity from solar panels. Instead of using distant energy sources, the generated electricity is used right away in the community for things like powering homes, charging EVs, and running local businesses. Excess energy is stored for later use or sale to the utility grid or other microgrids at times of highest demand and limited supply. As we discussed previously, monitoring and control systems help us use energy better and avoid waste. This short-cycle economy helps communities become more self-sufficient, resilient, and sustainable.

Moreover, a microgrid allows people to share energy-efficient electric appliances or tools, which can significantly reduce individual costs and environmental impact. For example, a community could share a fleet of EVs instead of each household owning one. Also, free, 100% electric laundromats could be available to everyone, freeing up space in individual homes or apartments if most people live in multi-family dwellings. Other examples include communal workshops, commercial kitchens for large gatherings and events, free central heating/cooling, etc. Essentially any activity or process that can be electrified is an option.

With the shared EV fleet, vehicles are used more efficiently and fewer would likely be needed than 1–2 per family. Promoting EVs through a shared model helps the community reduce its carbon footprint and be more environmentally sustainable. Shared EVs also save money by pooling purchase, insurance and maintenance costs, and leveraging economies of scale to buy in bulk. Community members save money and resources, and shared EVs bring people together to manage the fleet. This sort of resource sharing in a microgrid community would be beneficial for both the environment and the community.

This concept may face challenges in the US, where there is a strong cultural emphasis on personal independence and ownership. While a limited application of shared resources could work, it is more practical to focus first on sharing the microgrid's generation and storage assets, as these are intrinsic to the design of a microgrid. For example, communities can jointly invest in solar panels, inverters, and battery storage systems. This shared infrastructure can provide reliable and sustainable energy to all members of the community, ensuring that everyone benefits from the microgrid's capabilities and allowing people to utilize the free or low-cost power provided as they see fit. In the US, peripheral applications, such as the sharing of personal equipment and transport, would need to be addressed on a case-by-case basis and would be more attractive in urban or densely populated settings, where residential space is in short supply and having personal transport is less of an imperative. In Europe, where there is often a greater acceptance of communal living and shared resources, these concepts may be more readily adopted. For example, cities such as Amsterdam and Copenhagen have successfully implemented shared EV schemes and community energy projects, demonstrating the potential of such models [14].

In the US, a case study approach may be more effective in demonstrating the benefits and feasibility of shared resources within a microgrid community. By examining successful examples from Europe and adapting them to local contexts, communities can better understand how to implement these systems. This approach allows for tailored solutions that respect cultural preferences while promoting sustainability and resilience. Ultimately, the aim is to create a virtuous circle in which a community takes steps to reduce energy consumption, generate clean energy, and install modern energy management infrastructure. By focusing on shared microgrid generation and storage assets, communities can create a strong foundation for sustainability.

As the energy landscape changes, the concept of "The Commons" stands out. The Commons are resources that are owned and managed by the community [15]. Elinor Ostrom, who won the Nobel Prize in Economic Sciences in 2009, demonstrated that communities could manage common resources sustainably. Before Ostrom's work, the Commons were often seen as a problem, with the belief that overuse would lead to depletion or degradation. Ostrom's research showed that with proper governance structures, communities could effectively manage shared resources without depleting them.

A rural community with lots of sunlight could use it to make solar energy. Instead of each house having its own solar panels, the community has a solar farm. The community owns the solar panels, inverters, and energy storage systems. Governance structures ensure everyone has access and that the resources are managed sustainably. People in the community save money on their electricity. When people are involved in decisions, they feel a sense of ownership in the project. The community can use the solar farm to reduce its reliance on fossil fuels, lower its carbon emissions, and contribute to local environmental goals. The Commons help residents harness renewable energy, promote social equity, and strengthen community resilience in the face of energy challenges. The Commons also show us how the systems we rely on are connected and that we need to cooperate to keep them functioning optimally.

The Four Rs: Rethink, Reuse, Reduce, Recycle

The quest for autonomy, resilience, and sustainability is about more than just managing resources within a microgrid and achieving energy self-sufficiency. It also encompasses the broader principles of the circular economy [16], which emphasizes the continuous use of resources through the four Rs: Rethink, Reuse, Reduce and Recycle [17]. In this context, microgrids play a crucial role by enabling local production and consumption of energy, reducing reliance on external sources and minimizing waste. This approach has a direct impact on Scope 1, 2, and 3 CO_2 emissions. By integrating the four Rs into microgrid management, communities can increase their self-sufficiency and sustainability. For example, rethinking how we use energy and resources can lead to more efficient consumption patterns, while reusing and recycling materials can reduce the overall environmental impact. This holistic approach not only supports the stability and efficiency of microgrids but also fosters a more sustainable and resilient community.

Recycling is a good way to protect the environment by reducing waste. Buying a new chair made from recycled plastic is good. Recycling also uses energy to make new products, repackage them, and deliver them. This energy is already in the product. The energy needed for recycling is usually less than that needed for making new products. Recycling paper and cardboard uses less energy than making it from trees. Integrating recycling into energy systems helps us use energy more efficiently and be more sustainable. Microgrids can manage and distribute the energy used in recycling, reducing overall energy consumption and enhancing the resilience of the energy supply.

The second R is **Reduce**. Let us say you drive to work and the store every day. You might walk downtown to meet friends for coffee (if it is a 10-minute walk). Reducing energy use is important for a successful energy transformation. It is not just about the technology we use. It is also about changing how we act and how society changes. As energy consumers, we also have a role to play. This is where sharing is important. Most people have an electric drill at home. It's used for about 12 minutes in its whole life, as mentioned before. Why not make it available to others like books in a library?

The third R is **Reuse**. Recycle is balanced by reuse. You have an old chair you do not want. It can be recycled. It is a wooden chair made by a carpenter in the 1950s, and thus a vintage product that can be sold and reused. This is the best option for this product because it does not waste energy. If it is popular, new products should be designed to be upgraded. For example, a laptop is made of parts that can be replaced, so it does not need to be recycled.

A new R is **Rethink**. The concept of "rethinking consumption" entails the formulation of deliberate, informed choices prior to the acquisition of any given item. It encourages consumers to purchase goods and services produced and delivered within their local communities. This practice supports the circular economy by reducing the length of supply chains and the associated transportation emissions. To illustrate, the purchase of locally produced food items not only provides support to local farmers but also serves to reduce the carbon footprint associated with the transportation of goods over extended distances.

Consider a hypothetical community that decides to source its food from nearby farms. This not only ensures the produce is fresher but also reduces the emissions from transporting food across the country or even from overseas. Additionally, local farmers benefit from the direct support of their community, which contributes to the resilience of the local economy.

Microgrids fit into this model by producing and consuming energy locally. This local production not only ensures energy security but also allows the community to appreciate the value of the energy they use, fostering a sense of responsibility and sustainability. You can also buy a big electric car (EV) that uses a lot of energy to be produced. You might think you're doing good for the environment, but sometimes keeping your old car is better for the environment. So that is the fourth R, which is the last on the list.

Rethink is about thinking twice before buying a product. It is about changing how you consume and about being a wiser consumer.

Emission Reductions: Scopes 1, 2, and 3 [18]

As we have discussed, widespread adoption of microgrids based on renewable power also has the potential to lead to a significant reduction in greenhouse gas emissions, especially if fuel-burning appliances (e.g., stoves and ovens), heating systems, and transport are electrified. When we look at CO_2 emissions and their impact on sustainability, you will often hear about them being referred to as one of three scopes of emission.

Scope 1 direct emissions that a company or organization provides from sources that it directly owns or controls. This includes on-site combustion of fossil fuels in vehicles or equipment. Microgrids can help by producing local renewable energy that is used to power newly electrified (previously fossil fuel burning) equipment, appliances, and transport.

Examples for a Household

- Burning fuel (e.g., gasoline or diesel cars).
- Operating a gas-fired furnace or boiler to heat your home.
- Using propane or other fuels for cooking or heating.
- Generators fueled by diesel, gasoline, natural gas, or propane.

How to Reduce: These emissions are under your direct control, so you can take steps to reduce them by electrifying fuel-burning activities.

Scope 2 emissions are indirect emissions that result from the energy you buy and use. They are created at the source of energy production. Microgrids can help communities use less energy generated by natural gas, coal, fuel oil, or diesel burning power plants by using locally produced renewable energy.

Examples for a Household

- The electricity you use in your home, which can be produced from power plants that burn coal, natural gas, or fuel oil.
- If you use central heating in a condominium building, the furnace emissions associated with producing that heat.

How to Reduce: By generating carbon-free power onsite or successfully petitioning to electrify fuel-burning sources of power and heat, you can reduce these emissions.

Scope 3 is more complicated. It is indirect emissions that are not produced by your household but are associated with your activities up and down your value chain (suppliers, customers, etc.). Scope 3 emissions are all indirect emissions in an entity's value chain, including upstream and downstream activities. Our choices as consumers, like buying recycled products, sharing resources, and changing how we use things can help reduce Scope 3 emissions. Microgrids help primarily with Scope 1 and 2 emissions, but with Scope 3, the benefits are more difficult to quantify. For example, the reduction in consumption of fuels and services from oil and gas companies and electric utilities related to implementation of a microgrid would reduce emissions related to the ancillary operations (e.g., fuel transport), materials procurement, and support services for those companies, but by how much would be difficult to calculate.

Examples for a Household:

- **Product Purchases:** The emissions from producing, transporting, and disposing of the items you buy (e.g., clothing, electronics).

- **Transportation:** Emissions from using public transportation, flying, or shipping goods.
- **Waste Management:** Emissions from waste disposal.

How to Reduce: These emissions often make up a significant portion of your overall carbon footprint. It is difficult for individuals to exert much influence directly over these reductions. The most direct way is by purchasing products or services that come from companies that have committed to reducing their carbon footprints in their operations. This will not always be possible, but by educating oneself about the practices and commitments of the companies that have undertaken such initiatives, it can be possible to modify one's purchasing practices to have an impact on Scope 3 emissions.

By embracing the four Rs and using smart grid technologies, communities can address emissions from sources 1 to 3. This fosters a culture of sustainability and drives us towards a more environmentally conscious future.

Let's consider a simple example for a household:

Scenario: You live in a house with a gas furnace, drive a gas car, and use electricity from the grid.

Scope 1: Your gas furnace directly emits CO_2 when you heat your home. You can reduce this by improving insulation or switching to a more efficient or electric furnace.

Scope 2: The electricity you use comes from a mix of sources. If you switch to renewable energy (such as solar panels), you'll reduce Scope 2 emissions.

Scope 3: When you buy products (e.g., clothing, appliances), consider their environmental impact. Recycling and waste reduction also play a role.

For an industrial example, let's consider Holcim, a Swiss company specializing in building materials and aggregates. Holcim is a global leader in sustainable construction solutions, with a presence in around 60 countries and 60,000 employees. Holcim's global carbon dioxide (CO_2) emissions from 2020 to 2023, categorized by scope, show a notable trend [19].

The data reveal a significant trend in Holcim's emissions over the four-year period. In 2020, Holcim's total CO_2 emissions were approximately 139 million tonnes. This figure decreased slightly in 2021 to around 133 million tonnes. The downward trend continued in 2022, with emissions falling further to 130 million tonnes. However, in 2023, emissions increased slightly to 127 million tonnes.

The categorization of emissions into Scope 1, Scope 2, and Scope 3 helps to understand the sources of emissions and the effectiveness of Holcim's strategies for reducing its carbon footprint. In 2023, Scope 1 emissions, which are direct emissions from owned or controlled sources, amounted to 90 million tonnes. Scope 2 emissions, which are indirect emissions from the generation of purchased electricity, were 15 million tonnes. Scope 3 emissions, which include all other indirect emissions that occur in the value chain, totaled 22 million tonnes.

This breakdown highlights the areas where Holcim has made progress and where further improvements are needed. The data underscore Holcim's ongoing efforts to reduce its environmental impact, although the slight increase in 2023 suggests areas for further improvement. This information highlights the company's commitment to sustainability and the challenges it faces in achieving its environmental goals [20].

Social Marginal Cost (SMC) [21] is important in economics and environmental studies. It shows the cost to society of making one more of a good or service. This cost includes the direct expenses of the producer and the indirect costs or benefits that affect third parties.

SMC includes private marginal cost (PMC) and external costs. The PMC is the cost of making one more unit, including things like raw materials, labor, and other costs. External costs are costs not reflected in the market price, such as pollution, health impacts, and environmental degradation.

A factory that produces electricity using coal is an example. The factory's cost for producing one more unit of electricity might include the cost of coal, labor, and equipment maintenance. Burning coal releases pollutants into the air, causing respiratory problems and contributing to climate change. The public pays for health and environmental costs. The SMC is the PMC plus external costs. The SMC is the PMC plus external costs. If the PMC is $50 and external costs are $20, the SMC is $70.

Policymakers need to understand SMC so they can make rules and policies that show the real cost of making things. For example, a carbon tax can be put in place to make producers pay for the damage they do to the environment. This encourages them to make things in a cleaner way and helps reduce negative effects on others. SMC can help electricity markets price electricity fairly. For example, time-of-use rates can charge more during peak demand periods to encourage consumers to use less electricity during these times.

Global Regulatory Framework

You need to obtain a thorough understanding of the applicable laws, regulations and local ordinances if you want to set up a microgrid. This section looks at some of the international and national rules that support the development of decentralized energy systems and energy autonomy generally. Knowing these rules helps prosumers and communities work with policy goals, stay compliant and innovate.

International Regulations on Climate Change

The increasing frequency and severity of catastrophic extreme weather events has underlined the urgent need for a global energy transformation. International agreements, such as the Paris Agreement, are now putting pressure on all governments to act swiftly to mitigate climate change. The IEA has highlighted the need to triple renewable energy capacity by 2030 to meet the Paris Agreement targets.

The IEA's Global Roadmap to Net Zero by 2050, first published in May 2021 and updated ahead of the COP28 climate conference, outlines a comprehensive pathway for the energy sector to limit global warming to 1.5°C. This roadmap emphasizes the need for a massive scale-up of clean energy technologies within this decade, coupled with a doubling of progress in energy efficiency, to reduce demand for fossil fuels and achieve net-zero emissions relatively quickly.

However, the proliferation of microgrids, which are central to local energy production and resilience, faces significant regulatory challenges. These challenges include the need for harmonized international standards, supportive policies and significant investment. While regions such as the EU and the United States have made progress with comprehensive policies and financing mechanisms to support microgrids and renewable energy, the global landscape remains fragmented. The EU's Clean Energy for All Europeans package, the Renewable Energy Directive and the Energy Efficiency Directive, as well as the US Energy Independence and Security Act, the American Recovery and Reinvestment Act and the recent Inflation Reduction Act, illustrate the different approaches to promoting microgrid development.

To effectively deploy microgrids globally, there must be a concerted effort to align international regulations, incentivise investment, and ensure that technological advances are accessible and scalable across regions. This alignment is critical to creating a resilient, low-carbon energy system that can withstand the impacts of climate change and support sustainable development globally.

The EU is working to encourage the use of microgrids as part of the transformation to renewable energy and carbon neutrality. Here are some key regulations and policies:

- **The 2019 Clean Energy for All Europeans Package** [22]**:** The 2019 Clean Energy for All Europeans Package is a key part of the EU's strategy to achieve a secure, competitive, and sustainable energy system. It creates a complete set of laws to support the Energy Union. Each EU country must develop a 10-year plan for 2021–2030 that shows how it will help the EU reach its targets for energy efficiency, renewable energy, and greenhouse gas reduction. These plans also look to the future, with a vision that extends up to 2050.
- **The EU Renewable Energy Directive** [23]**:** Sets renewable energy targets for EU countries. The Renewable Energy Directive 2018 is a EU law that requires 42.5% of EU energy to be renewable by 2030. The target is shared among member states.
- **The EU Energy Efficiency Directive** [24]**:** Aims to improve energy efficiency. EU countries must save energy over the next 10 years. They must save 0.8% of final energy consumption in 2021–2023, 1.3% in 2024–2025, 1.5% in 2026–2027, and 1.9% in 2028–2030.
- **The European Green Deal** [25]**:** It is the EU's plan to make the economy sustainable and achieve climate neutrality by 2050. The EU wants to use more renewable energy and microgrids to reduce carbon emissions. Programs fund research, innovation, and smart grid and renewable energy projects.

- **Horizon Europe** [26]**:** It is the EU's research and innovation program. It funds research on climate change and energy. Smart grid projects can get funding under Horizon Europe.

The EU is actively working to promote the use of smart grids as part of its transformation to renewable energy and carbon neutrality. Recognizing the essential role of smart grids in improving energy efficiency, reliability, and resilience, the EU has implemented several key regulations and policies to support this transformation. The 2019 Clean Energy Package for All Europeans is a cornerstone of the EU's energy transformation plan, establishing a robust framework for the Energy Union. The package requires each EU country to draw up a 10-year energy and climate plan for 2021–2030, setting out how it will meet its Energy Union targets and looking ahead to 2050.

The EU's Renewable Energy Directive sets ambitious renewable energy targets for member states, requiring 42.5% of the EU's energy to come from renewable sources by 2030. The EU's Energy Efficiency Directive aims to improve energy efficiency across the bloc, requiring countries to make incremental energy savings over the next decade, with specific targets for each period.

The European Green Deal is the EU's comprehensive plan to make the economy sustainable and carbon neutral by 2050, promoting the use of renewable energy and smart grids to reduce carbon emissions. This plan includes funding for research, innovation, and projects related to smart grids and renewable energy through programs such as Horizon Europe, the EU's research and innovation program.

Together, these measures create a policy environment that promotes the deployment of smart grids and the transformation to a low-carbon energy system by encouraging investment in smart grid infrastructure, increasing the use of renewable energy sources, and improving energy efficiency. These efforts are crucial for the EU to meet its climate and energy targets and to lead the global transformation to a sustainable energy future.

On the other side of the Atlantic Ocean, the United States is actively working to modernize its energy infrastructure and has passed several laws to promote the use of smart grids as part of its transformation to renewable energy and carbon neutrality. Recognizing the critical role of smart grids in improving energy efficiency, reliability, and resilience, the US has implemented the following key regulations and policies to support this transformation.

- **Energy Independence and Security Act (EISA) of 2007** [27]**:** This act established a national policy for grid modernization and provided incentives for stakeholders to invest in smart grid initiatives. Title XIII of the EISA specifically focuses on the development of a smart grid, promoting the use of advanced technologies to improve the reliability, efficiency, and security of the electricity grid.
- **Bipartisan Infrastructure Law (BIL) of 2021** [28]**:** This law includes provisions for upgrading the nation's power infrastructure, supporting the deployment of smart grid technologies, and enhancing grid resilience. It

also provides funding for research and development of advanced energy technologies.

- **Inflation Reduction Act (IRA) of 2022** [29]**:** This act includes significant investments in clean energy technologies, providing production and investment tax credits for renewable energy generation, energy storage, and EVs. The IRA aims to reduce greenhouse gas emissions and promote the transformation to a low-carbon economy. Many of the provisions in the IRA will be reduced or repealed by legislation passed by US Congress in 2025 quizzically referred to as the "Big Beautiful Bill".
- **Federal Energy Regulatory Commission (FERC) Order 2222** [30]**:** This order enables DERs to participate in wholesale energy markets, promoting the integration of renewable energy sources and enhancing grid flexibility. It supports the development of more decentralized and resilient energy systems.

Together, these policies create a supportive environment for the deployment of smart grid technologies and the transformation to renewable energy in the US, helping to meet the nation's energy security goals.

The comprehensive regulatory and policy frameworks established by both the EU and the US are key to driving the adoption of energy self-sufficiency and microgrids. As a result, these schemes are expected to significantly increase the deployment of microgrids, enabling communities to achieve greater energy self-sufficiency and resilience.

In conclusion, while regulatory frameworks and policies in the EU and the US are key to promoting energy self-sufficiency and the deployment of microgrids, several challenges remain. These include regulatory complexity and fragmentation, high upfront costs, technology integration issues, grid stability and security concerns, market and policy uncertainty, public acceptance and behavior change, infrastructure upgrading needs, and the need for effective stakeholder coordination. Overcoming these barriers is essential to realize the full potential of microgrids in creating a resilient, sustainable, and decentralized energy system.

What are Sustainable Development Goals (SDGs) and How Are They Important for the Dissemination of Smart Grids?

The 17 SDGs [31] set by the United Nations are becoming more and more widely adopted. They are interesting, since they are based on the three original pillars of the sustainable development model: environment, economics, and society. Below is the list of all the SDGs:

1. No poverty (SDG 1);
2. Zero hunger (SDG 2);
3. Good health and well-being (SDG 3);
4. Quality education (SDG 4);
5. Gender equality (SDG 5);
6. Clean water and sanitation (SDG 6);
7. Affordable and clean energy (SDG 7);

8. Decent work and economic growth (SDG 8);
9. Industry, innovation and infrastructure (SDG 9);
10. Reduced inequalities (SDG 10);
11. Sustainable cities and communities (SDG 11);
12. Responsible consumption and production (SDG 12);
13. Climate action (SDG 13);
14. Life below water (SDG 14);
15. Life on land (SDG 15);
16. Peace, justice, and strong institutions (SDG 16); and
17. Partnerships for the goals (SDG 17).

They are often visually presented with this poster presented in Figure 4.3.

The UN's SDGs address global challenges across environmental, economic, and social dimensions. Microgrids and self-sufficiency help with several SDGs, especially those addressing energy, the environment, growth, and equality. Here is how microgrids and self-sufficiency contribute to specific SDGs:

- **SDG 7: Affordable and Clean Energy.** Microgrids help achieve SDG 7 by integrating renewable energy sources, improving energy efficiency, and ensuring reliable and affordable electricity access. Microgrids help create cleaner and more sustainable energy systems, which is important for achieving universal access to affordable and clean energy.
- **SDG 9: Industry, Innovation, and Infrastructure.** Microgrids help achieve SDG 9 by encouraging new energy technologies. Microgrid investments help countries become more resilient and efficient in their energy systems, which helps the economy grow, create jobs, and develop industry.
- **SDG 11: Sustainable Cities and Communities.** Microgrids help create sustainable cities and communities that use energy wisely and protect the environment. Microgrids help create liveable, inclusive, and sustainable urban environments by using energy more efficiently, reducing emissions, and making cities more resilient to climate change.
- **SDG 13: Climate Action.** Microgrids help advance SDG 13 by integrating renewable energy, reducing greenhouse gas emissions, and enhancing climate resilience. Demand response, grid optimization, and DERs help mitigate climate change impacts and support global climate targets.
- **SDG 17: Partnerships for the Goals.** Microgrids help people from different groups work together to achieve sustainable development. By sharing knowledge, transferring technology, and building skills, microgrids help people from different groups work together to achieve the SDGs.

This section examined some of the international and national rules that support the development of energy systems. Knowing these rules helps prosumers and communities strive toward political goals, stay within the law and come up with new ideas.

As one can see from what we just described, the development and use of microgrids is influenced by a complicated set of rules and politics that are different

Fig. 4.3 The 17 SDGs of the UN. [32]

in each country and even in each region. It is important to have good rules to encourage people to invest in microgrids, make sure that microgrids can work with other systems, and make sure that everyone has access to energy. Germany has been a leader in developing renewable energy and microgrids. The German Renewable Energy Act (EEG) created a good environment for investment in renewable energy and increased the growth of the microgrid market. Denmark has an ambitious energy policy that aims to achieve carbon neutrality by 2050. It also has a good system for microgrids, especially in rural and island areas. The US has a more varied system, with different policies and incentives in each state. But some states, like California and New York, have programs to support microgrid development.

Regulators are generally trying to make it easier to connect microgrids to the main grid, which will reduce costs and delays. The prices of electricity can change a lot, and this can encourage people to use their energy in the best way and make the most of microgrids. Regulators are looking into the idea of creating energy markets where prosumers can trade energy with each other. The rules and future trends have an effect on how microgrids are run and how change is managed. The next section talks about the best ways to manage microgrid operations and move to energy models that use less resources.

Operational Strategies and Change Management

Good strategies and change management are important for adopting microgrid systems. This section outlines the best ways to manage microgrid operations. Community engagement, continuous improvement and adaptive management are key to the long-term success of any energy community.

Off-grid and On-grid Differences

The debate between off-grid and on-grid solutions in the evolution of microgrids revolves around the optimal balance between energy independence and grid connectivity. Let us explore both approaches.

Off-grid microgrids operate independently of the centralized utility grid, relying solely on localized energy generation and storage. These systems are often deployed in remote or rural areas where grid connectivity is unavailable, limited or unreliable. For example, a remote village in the mountains lacks access to the centralized electricity grid. To meet their energy needs, the community installs solar panels, wind turbines, and battery storage systems to create an off-grid microgrid. The system is managed through monitoring and control technologies, optimizing energy usage and ensuring reliable power supply to residents.

On-grid microgrids are connected to the larger grid network. They can access the grid for backup power during periods of high demand or reduced energy generation. On-grid microgrids are more suitable for suburban and urban areas and commercial developments. For instance, a university campus uses solar panels and energy storage to create an on-grid microgrid. During sunny days, the microgrid generates excess solar energy, which is exported to the utility grid. During cloudy days or peak demand periods, the campus draws power from the grid when microgrid storage is inadequate to meet demand.

Generally, off-grid systems require more upfront investment compared with on-grid microgrids. However, on-grid systems may have additional charges.

- **Reliability:** Off-grid systems can be reliable, but their reliability requires a source of non-intermittent power generation. While on-grid systems can lose power during grid outages, an islandable on-grid microgrid has a very low probability of losing power completely. Conversely, an off-grid system could be knocked offline if one of its components fails.
- **Scalability:** On-grid systems can easily change how much energy they use, while off-grid systems may require more investment to expand capacity.

The choice between off-grid and on-grid micro smart grids depends on things like location, energy needs, budget, and long-term goals as well as regional and national regulations. Some projects may benefit from a mix of off-grid and on-grid solutions, especially when certain critical systems need to be shielded from external threats, whether physical, phygital or purely digital [33].

Change Management

You want to become a prosumer to be more energy independent. Why? You may care about the environment. Or it might be because the grid is unreliable. Or you think you can save money on your electricity bills. But becoming a prosumer means changing how you use electricity. You will now produce your own energy and rely on others in the energy community. This is a process of change management. It is about how you act, and this is especially important if you're in a group. Such a change takes time. It will take at least two to three years to really change how you produce and consume energy. You have to be persistent, patient, and committed.

To ensure the successful adoption and operation of microgrid systems, it is crucial to have effective operational strategies and proactive change management in place. This section emphasizes the importance of community engagement, continuous improvement, and adaptive management in building a strong and resilient energy community.

Two-thirds of change management projects fail. From an energy transformation perspective, change is seen as a continuous process. There are group and organizational dynamics in which complex forces are at work. This fits well with Kurt Lewin's model of change (1945), which says that new prosumers should be aware of two types of resisting forces: habits and social customs; and strong forces that can break habits, challenge interests, or disrupt group customs, which can promote radical change.

Kurt Lewin's model of change, often referred to as the unfreeze-change-freeze model, provides a framework for understanding and managing organizational change. In the context of energy communities and the transformation towards self-sufficiency, this model emphasizes the importance of preparing individuals and organizations for change before implementing radical transformations. Here is how the model applies:

Unfreeze (Preparation for Change)

Before embarking on a radical change such as becoming energy independent, it is essential to "unfreeze" existing attitudes, behaviors, and systems within the energy community. This involves creating awareness of the need for change and addressing any resistance or inertia that may hinder progress. Prosumers must be willing to challenge existing habits and social norms related to energy consumption and reliance on centralized utilities. For instance, before introducing the concept of self-sufficiency to an energy community, educational workshops and outreach programs can be organized to raise awareness about the benefits of renewable energy, energy conservation practices, and the potential for decentralized energy systems. Engaging community members in discussions about their energy needs, concerns, and aspirations helps unfreeze existing mindsets and attitudes towards change.

Change (Implementation of Radical Change)

Once the energy community is ready and receptive to change, the next step is to implement radical changes aimed at achieving self-sufficiency and adopting sustainable energy practices. This may involve deploying smart grid technologies, investing in renewable energy infrastructure, and fostering collaborative energy management initiatives among prosumers. For example, with the support and participation of community members, the energy community installs solar panels, wind turbines, and battery storage systems to establish a microgrid. Prosumers actively engage in energy monitoring, demand response programs, and peer-to-peer energy trading, enabling the community to reduce reliance on external energy sources and achieve greater self-sufficiency.

Freeze (Solidifying New Practices)

Finally, the "freeze" stage involves solidifying and institutionalizing the changes introduced during the transformation process. This entails embedding new practices, norms, and behaviors into the fabric of the energy community to ensure long-term sustainability and resilience. Continuous monitoring, evaluation, and adaptation are essential to maintain momentum and address emerging challenges. For example, the energy community establishes governance structures, policies, and incentives to support ongoing collaboration, innovation, and growth in self-sufficiency efforts. Regular community meetings, workshops, and training sessions reinforce the importance of sustainability and empower prosumers to take ownership of their energy future.

By following the unfreeze-change-freeze model, energy communities can navigate the complexities of change management and successfully become self-sufficient. By addressing resistance, fostering collaboration, and embracing innovation, prosumers can lead the way towards a more sustainable and resilient energy future.

Figure 4.4 shows that our assessment (Fragnière et al., 2023) found more restraining than driving forces. This shows that today, according to Kurt Lewin's change management theory, the risk of success for a microgrid is low. To increase the probability of success, you must either remove the obstacles or add more advantages. This analysis is qualitative, but it shows that we need to do more to get people to accept microgrids.

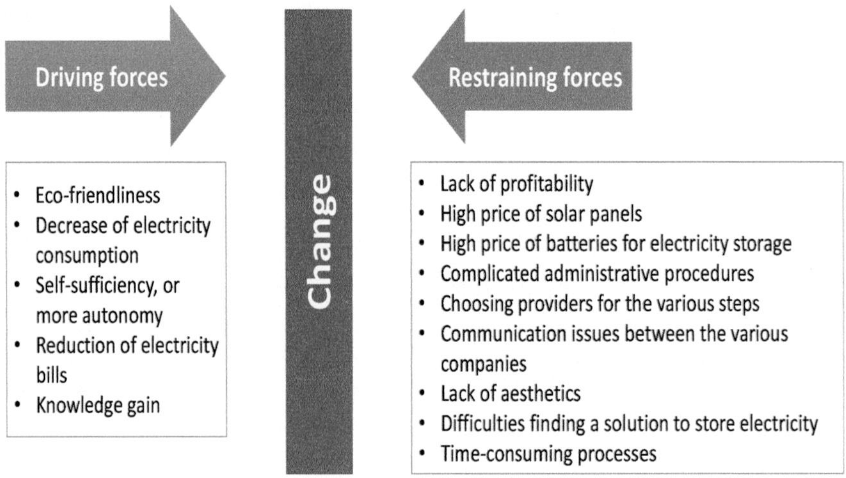

Fig. 4.4 An application of the Force Analysis as proposed by Kurt Lewin [34].

You cannot start a change project until you have analyzed the forces and found a positive ratio of driving to restraining forces.

The Four D's of Microgrid Energy Communities

As the world grapples with the challenges of climate change and the need for sustainable energy solutions, the concepts of decarbonization, decentralization, digitalization, and democratization have emerged as critical elements in the evolution of energy systems [35]. These concepts are particularly relevant when discussing microgrid energy communities and the pursuit of energy self-sufficiency. Microgrids, as localized energy systems that operate independently or in conjunction with the larger grid, offer a promising path forward. By integrating these four key principles, microgrid energy communities can lead the way to a more sustainable and resilient energy future.

Decarbonization is the first and most critical step in creating a sustainable energy system. It involves reducing or eliminating carbon emissions by switching from fossil fuel to renewable energy sources such as solar, wind, and hydropower. In the microgrid context, decarbonization means prioritizing the use of these clean energy sources to minimize the environmental impact of energy production. This not only helps combat climate change but also reduces the community's dependence on external energy supplies and promotes greater energy independence.

Decentralization is about moving away from large, centralized power plants that generate electricity and distribute it over long distances. Instead, it focuses on creating smaller, localized energy systems that generate power close to where it is consumed. For microgrid energy communities, decentralization means that energy is generated, stored, and used within the community itself, reducing transmission losses and increasing energy reliability. This is especially important in remote or rural areas where access to the central grid may be limited or unreliable.

Digitalization refers to the incorporation of digital technologies to manage and optimize energy systems. In a microgrid, this can include the use of smart meters, sensors, and advanced software that monitor energy production and consumption in real time. Digitization enables microgrids to be more efficient by ensuring that energy supply precisely matches demand, reducing waste, and facilitating the integration of variable renewable energy sources. In addition, digitized systems can provide valuable data to help communities make informed decisions about their energy use and future infrastructure investments.

Democratizing energy is about empowering individuals and communities to take control of their energy future. It means giving people a greater say in how their energy is produced, distributed, and consumed. In microgrid energy communities, democratization takes the form of increased community involvement in energy planning and decision-making. Residents can participate in the management of their energy resources, choose how to source their electricity, and even become energy producers themselves, for example by installing rooftop solar panels. The result is a more engaged and empowered community where energy is not just a commodity but a shared resource that benefits everyone.

Managing a Smart Grid Like an Enterprise

It is critical for our Energy Community to define a new name and tagline that embodies our identity as a forward-looking and sustainable entity. This requires strategic thinking that considers the triad of mission, vision, and values. By clearly defining purpose, values, and identity, the community can communicate with its prosumers in a consistent and unified manner.

Defining the mission-vision-values triad is fundamental to an energy community. It allows it to define its purpose (mission), set the long-term goal (vision), and determine the guiding principles that will guide its actions and decisions (values). In short, strategic thinking begins with aspirations and future direction (vision), continues with what must be accomplished to achieve that vision (mission), and concludes with the values that will guide operations and actions.

When the energy community has a clear understanding of its mission, vision and values, it can make coherent decisions and work effectively to achieve its goals. It also helps community members understand their roles and work together in a common direction. Finally, it enables the community to effectively communicate its goals and values to its stakeholders.

Examples of Vision, Mission, and Values for an Energy Community

Example 1

- **Vision:** To be a leading model for sustainable living, harnessing renewable energy for a greener future.
- **Mission:** To empower our community of prosumers by providing a platform for sustainable energy practices, promoting energy independence, and encouraging the use of renewable energy sources.
- **Values:** Sustainability, innovation, collaboration, empowerment, and responsibility.

Example 2

The following is an example of a vision, mission, and values statement for an energy community with a goal of safety and reliability.

- **Vision:** To create a safe and reliable energy community that prioritizes the well-being of its members and the environment, setting the standard for sustainable living and energy resilience.
- **Mission:** It is our mission to provide secure and uninterrupted energy services to our members, ensuring their safety, comfort, and peace of mind.
- **Values:**
  - **Safety:** We prioritize the safety and well-being of our members, employees, and the surrounding community in all aspects of our operations and decision-making.
  - **Reliability:** We strive to deliver reliable and resilient energy services that meet the needs and expectations of our members, maintaining high standards of performance and accountability.
  - **Collaboration:** We believe in the power of collaboration and partnership, working closely with our members, stakeholders, and industry partners to enhance safety, reliability, and innovation in our energy community.
  - **Continuous Improvement:** We are committed to continuous improvement and innovation, proactively identifying opportunities to enhance safety, reliability, and efficiency in our energy systems and practices. It is our responsibility to protect and preserve the environment for future generations. We integrate sustainable practices and renewable energy solutions into our operations and initiatives. Our energy community aspires to create a safe, reliable, and sustainable energy ecosystem that serves as a model for others to emulate. We can build a brighter and more resilient energy future for all through collaboration, innovation, and a shared commitment to safety and reliability.

Conclusion

When it comes to operational strategies, there are several key areas to focus on. Real-time monitoring and control is essential, as it allows for the tracking of microgrid performance in terms of energy production, consumption, and battery status. By implementing advanced monitoring systems, operators can make informed decisions to optimize the system's efficiency.

Another important strategy is optimizing the use of renewable energy sources. Using intelligent algorithms, microgrids can maximize the utilization of renewable energy and minimize reliance on traditional energy sources. This not only reduces the environmental impact but also contributes to the long-term sustainability of the system.

Demand management is also a key strategy to consider. By implementing demand response programs, consumers can be encouraged to reduce their energy consumption during peak periods. This not only helps to balance the grid but also reduces overall energy costs.

Preventive maintenance is crucial to ensure the reliability and longevity of microgrid equipment. Establishing a preventive maintenance program allows for the identification and resolution of potential issues before they become major problems, thus ensuring the continued operation of the system.

In terms of change management, the transition to a decentralized energy model based on microgrids requires significant changes for individuals, communities, and organizations. Therefore, effective change management is vital to ensure successful adoption and stakeholder satisfaction. Clear communication and stakeholder engagement are necessary to explain the benefits of microgrids and involve stakeholders in the planning and implementation process. Training and skills development for microgrid operators and users are also important aspects of change management. Providing the necessary training allows individuals to understand the technical and operational aspects of the system, empowering them to effectively utilize and maintain the microgrid.

Continuously monitoring the performance of the microgrid and evaluating its effectiveness is crucial. This allows for any necessary adjustments to operational strategies and change management programs to be made, ensuring the long-term success of the microgrid system.

Overall, effective operational strategies and successful change management are crucial for the adoption and operation of microgrid systems. These strategies not only contribute to the building of a strong and resilient energy community but also pave the way for new opportunities and business models in the evolving energy landscape.

Future Opportunities and Business Models

Microgrids offer the prospects of developing new business models and spurring economic growth. This section looks at new trends, markets, and business strategies for microgrids. By taking advantage of these opportunities, prosumers and communities can create sustainable and profitable energy solutions. This fits with the chapter's vision of a dynamic and forward-thinking energy landscape.

Gamification to Improve Our Energy Behavioral Consumption [36]

Climate change is an increasingly concerning issue and it seems natural disasters exacerbated by the effects of climate change are being reported almost daily. To mitigate the impacts of climate change, many initiatives are being launched by governments, industries and communities around the world, at the international, national, regional, and local levels. Despite this groundswell of activity and awareness, attempts at changing environmentally harmful behaviors still face many hurdles. In the last few years, gamification and serious games have been identified as potential methods of expanding awareness of the deleterious effects of human activity on climate change and spurring action to mitigate those impacts.

Games are generally used for learning and developing useful skills. This has given rise to the terms "serious games" and "gamification". This method of learning has shown positive results engaging and motivating people to achieve a goal.

Gamification is a way of engaging users in real-world activities by using game design principles in a non-gaming context. Serious games are simulations with educational goals. They include rules, choices, and challenges. They can prepare people for real-world tasks by facing challenges in the game. Studies have shown that scholars are very interested in using games for sustainable development. The research also demonstrates that games are effective educational tools. A growing number of studies, including detailed reviews, show an increasing interest in serious games for teaching people about the environment. These games have the potential to get learners interested in complex environmental issues [37]. In addition, studies show that games and gamified approaches are effective in teaching people about the environment, encouraging positive attitudes and behaviors [38]. Gamification can result in longer-term psychological involvement, unlike other approaches to behavior change. By gamifying the challenges and potential positive results of implementing and managing microgrids, teaches people how to do things in a responsible and sustainable way.

The social aspect is important in the game. People want to be validated. The game could show people that they are not alone in trying to make a difference. Showing other players' efforts could convince them they are all involved. People also want to feel like they are part of a group and that they are contributing by showing off their efforts and getting social recognition. The game mechanics could be a leaderboard, competition, or community.

It is also important to encourage people to take action. Studies show that people don't always need tangible rewards to be motivated—sometimes, simply acknowledging their efforts is enough. The game could provide feedback and recognition, showing how individual actions affect others and emphasizing the positive impact on the community. This can make people feel valued and appreciate what they're doing, which helps motivate them to keep going. For example, players could be shown how their behavior reduced energy use or helped other people. People are more likely to keep acting sustainably if they see that their efforts matter.

Finally, governments can support gamification in many ways. They can invest in things like recycling facilities that encourage sustainable behaviors. They can also use digital tools to track what people are doing and provide them with recognition or consequences depending on how much they take part. This helps make people more likely to carry on with their eco-friendly actions and shows that the government is serious about climate action.

If the government is involved in the climate change gamification effort, it will give the effort more credibility and likely draw more people to follow the initiative. The game being proposed and developed by the government can build trust with the citizens as it shows that the government cares about the environment and is willing to act. This can create momentum and lead citizens to do the same. Moreover, it ensures that people cannot cheat as it is monitored by a higher institution.

For example, ECOTREE [39] is a game where you save energy at home. It is a game that teaches people to save energy in small ways. You can play the game on your phone or computer. People typically do not actively think about how much energy they use on a regular basis. It helps them understand how much energy they use and how to use less. Saving energy helps the environment and also saves people money.

In the game, players take care of virtual trees called Treenergy (see Figure 4.5). The game is like Tamagotchi, a virtual pet game. It runs on a mobile app that you can download to your phone or tablet. The goal is to keep the trees healthy and happy. This is affected by how much energy people use at home. The happier and healthier their Treenergy, the better they are at managing their consumption.

Fig. 4.5 A few snapshots of the ECOTREE app [40].

It should be noted that there are many other gaming applications that have been developed with the objective of reducing energy consumption. The following list includes some examples of app-based games specifically focused on teaching energy consumption and efficiency:

- **Energy Transformers** [41]**:** This educational game lets players explore renewable energy solutions and their impact on health and the environment. Players team up with a character named Terra to reduce climate pollution and increase renewable energy use.
- **Energy Efficiency Calculator** [42]**:** Created for Georgia Power's Learning Power program, this interactive tool helps users learn about energy efficiency through fun activities.
- **Think! Energy Games** [43]**:** These interactive games from Consumers Energy include challenges like word searches and puzzles that teach energy efficiency and ways to save energy at home.
- **Let's Talk Energy Games** [44]**:** This collection of online games helps players understand energy use and climate change. Activities include defeating energy hogs in your home and building power-generating stations.

New, Interesting Jobs for the Younger Generations [45]

The outlook for employment opportunities in the US in the context of energy decentralization is highly promising. The US Government has made significant commitments to address climate change and accelerate the transformation to renewable energy, which is expected to drive job creation across various sectors of the economy. Several factors contribute to the positive outlook for jobs in the US.

The US Government has proposed a series of ambitious plans to invest in resilient, decentralized, and low carbon energy infrastructure. These include the construction of renewable energy generation facilities, the installation of EV charging stations, the construction of energy-efficient buildings, and the modernization of grid systems. It is anticipated that these investments will result in the creation of an important number of new jobs in various sectors, including construction, manufacturing, engineering, and related fields [46].

The development of microgrids as part of the energy transformation is expected to create new job opportunities for younger generations. Potential roles include:

- **Microgrid Systems Engineer:** A microgrid systems engineer is responsible for the design and development of the architectural framework of smart microgrids. This involves the integration of renewable energy sources, storage systems, and demand response technologies in a manner that optimizes energy management, thereby enhancing grid stability and sustainability.
- **Storage Specialist:** With the increasing variety of energy storage types and form factors, there will be a growing demand for specialists who can manage these various technologies and systems and integrate them into microgrids.
- **Renewable Energy Technician:** Professionals who install and maintain renewable energy infrastructure, such as solar panels and electric heat pumps, within the microgrid.
- **Data Analysts and AI Specialists:** They will be needed to analyze and optimize the performance of these systems. Microgrids generate a significant

amount of data, requiring professionals to analyze it for energy production and consumption optimization. Additionally, they can develop AI models to automate and improve microgrid operations.

- **Community Energy Managers:** Specialists will manage the energy community, ensuring all members are engaged and committed to the project. They would oversee the operation of the smart microgrid and coordinate with members for energy production and consumption.
- **Energy Policy Advocates:** They will be required to work to influence energy policies and regulations to support the development and operation of smart microgrids and energy communities.

These roles contribute to the energy transformation and offer exciting opportunities for the younger generation to be at the forefront of this change.

New Business Models

The development of microgrids is creating a new wave of business models in the electricity sector. These models are designed to make clean, reliable energy more accessible and affordable, particularly for communities and businesses that lack the resources to invest heavily upfront. Some important examples of this are microgrid-as-a-service (MaaS), pay-as-you-go (PAYG) systems, mixed ownership structures, and new models like virtual power plants (VPPs) [47]. These approaches show a bigger change towards more flexible, scalable, and collaborative solutions for distributed energy systems. At the same time, the declining costs of installing and operating microgrids, increases in how much money they can save or generate and the availability of financial incentives are making the business environment more hospitable and diverse. Energy communities and financing partnerships are becoming more important. They help make the transition to energy independence more inclusive and achievable.

Microgrid-as-a-Service (MaaS)

Description: MaaS allows businesses and communities to access microgrid services without upfront capital investment. Service providers own, operate, and maintain the microgrid infrastructure, offering reliable power supply and energy management solutions on a subscription basis.

Example: A rural community partners with a renewable energy company to deploy a solar-powered microgrid. The company installs, operates, and maintains the microgrid infrastructure, providing electricity to residents and businesses on a monthly subscription plan [48].

Pay-As-You-Go (PAYG) Models

Description: PAYG models involve microgrid developers owning and financing the infrastructure for subscribing customers or electricity buyers. Customers pay for energy as they use it, with flexible payment options and access to reliable power.

Example: A remote village in a developing country lacks reliable electricity. A microgrid developer installs a solar-powered microgrid and battery storage system, offering residents the option to purchase energy credits through mobile payments, ensuring access to affordable and sustainable energy services [49].

Mixed Ownership Models

Description: Utilities, technology providers, and third parties work together to finance and operate microgrid projects. This reduces risks, lowers costs, and speeds up the deployment of microgrid infrastructure.

Example: A utility company works with a renewable energy developer and a community organization to set up a grid-connected microgrid in an urban neighborhood. The utility provides grid integration services, the developer supplies renewable energy technologies, and the community organization facilitates community engagement and participation [50].

Virtual Power Plants (VPPs)

Description: VPPs aggregate multiple microgrids and DERs to operate similarly to a single power plant. This model enhances grid stability and allows for better integration and optimization of renewable energy sources.

Example: A city integrates several microgrids and DERs into a VPP, which is managed by an AI-based control system. This setup provides grid services such as frequency regulation and demand response, improving overall grid reliability and efficiency [51].

These innovative business models for microgrid dissemination demonstrate the diverse approaches available to achieve profitability while promoting sustainability, resilience, and energy independence. The deployment of microgrids, which leverage technology, collaboration, and financing mechanisms can unlock long-term value for stakeholders and contribute to the transformation towards a more sustainable energy future.

Reference Energy System (RES)

The RES is a process that begins with the extraction of raw materials and culminates in the production of final energy carriers (see Figure 4.6, for example). It is crucial to consider the complete energy conversion cycle of the RES to accurately assess the efficiency of power generation, as there is inevitably a degree of waste at each transfer or transformation point. Typically, coal is extracted from an underground mine and then burned to generate electricity. Subsequently, the electricity is conveyed through the grid and subjected to voltage and current conversions at multiple points before reaching the consumer's outlet. The overall efficiency of the process of converting coal into electricity is approximately 33%. This means that two-thirds of the energy content of the coal is lost during the conversion process. Furthermore, coal-fired power plants emit approximately 2,257 pounds of CO_2 MWh of electricity generated [52].

The challenge of stability is really the grid's major challenge. Given the limited and expensive means of storing electricity, it is currently optimal to use rather than store it.

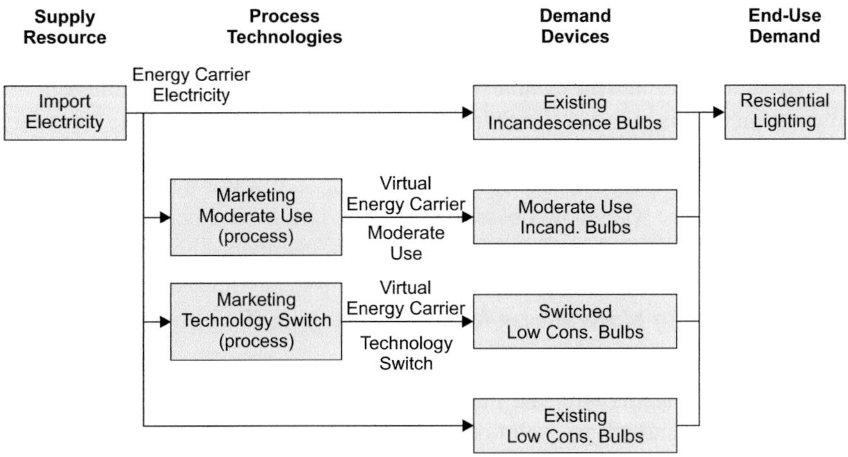

Fig. 4.6 An example of an RES (*Source:* Authors).

Modifying the configuration of the RES is a challenging endeavor due to the deeply entrenched nature of household energy behavior. Up to 95% of household energy behavior can be classified as a form of habitual behavior [53, 54]. This indicates that most energy consumption patterns are automatic and routine, rendering them resistant to alteration.

To disrupt this habitual cycle, it is imperative to eliminate the incentives that perpetuate these behaviors, to educate consumers about their habits, and to empower them to regulate the consequences of their actions. For instance, the provision of real-time feedback on energy usage can assist consumers to become more aware of their consumption patterns, thereby encouraging more efficient behavior. Furthermore, the implementation of demand response programs can provide an incentive for consumers to reduce or shift their energy usage during peak periods, thereby promoting sustainable energy practices.

For example, dynamic pricing and time-of-use tariffs can encourage consumers to use less energy when it is in demand and more when it is not. Another strategy is to raise awareness of energy consumption habits. Providing feedback on energy usage can help consumers identify areas of high usage. It is similarly crucial to empower consumers to regulate their energy consumption. The implementation of tools and technologies, including programmable thermostats, smart appliances, and home energy management systems, empowers consumers to more effectively regulate their energy consumption. Such technologies can be programmed to operate during off-peak hours or adjust settings based on occupancy, thereby aligning energy usage with consumers' preferences and priorities. Transformative change in household energy usage is vital for the transition towards a consumer-centric energy system. By eliminating incentives for wasteful practices, enhancing awareness of energy consumption, and empowering consumers with greater control, we can enhance the sustainability of our homes. This shift not only benefits individual households but also contributes to broader environmental goals by reducing overall energy consumption and greenhouse gas emissions.

By addressing both the technical aspects of the RES and the behavioral patterns of consumers, it is possible to enhance the efficiency and sustainability of energy systems. This comprehensive strategy not only enhances energy management practices but also facilitates a cultural transformation towards a more sustainable lifestyle. For those wishing to observe a real-time energy dashboard in action, the following website may be of interest: https://www.energydashboard.co.uk/live.

New, Surprising Motivations for Energy Self-sufficiency

A recent paper [55] looks at how people in the US are using private solutions to make up for problems with the public electricity grid. These solutions include backup batteries and off-grid solutions. People with more money are more likely to use these solutions which shows that not everyone has the same access to solutions for grid instability. The integration of DERs into existing grid infrastructure represents a major change towards more resilient and sustainable energy systems. Recent research shows that using local microgrids can make energy security better and reduce greenhouse gas emissions [56]. These local energy networks help communities keep the lights on during power outages and make the most of renewable energy sources using advanced management systems. What's more, when microgrids are used, they can do more than just save people money. They can also help to reduce the pressure on the main grid and create chances for people to trade energy with each other in their communities.

While only a fraction of households currently adopt private substitutes to utility grid power supply, their availability can benefit the majority. However, concerns about equity arise, as these disparities in adoption rates may exacerbate social inequalities. But it also presents opportunities to think differently about energy resilience and efficiency. To maximize the opportunities for microgrids to inter-operate and proliferate, it greatly helps to ensure everyone has access to reliable energy solutions either independently or through their community leadership.

The search for new opportunities and economic models shows how microgrids can transform the energy landscape. By taking advantage of these opportunities, prosumers and communities can create sustainable and profitable energy solutions, which matches the chapter's vision of a dynamic and forward-looking energy landscape. The growth of microgrids and the move towards more decentralised energy systems creates new opportunities and economic models. The next section talks about the best ways to manage microgrid operations and move towards more sustainable energy models.

Transformation: A Pragmatic and Visionary Approach

The necessary transformation of the energy infrastructures in developed nations is one of the major challenges of the 21st century. However, for it to be successful, it must not only be sustainable but also economically beneficial. It must not compromise the current standard of living but instead significantly improve overall quality of life. This section concludes the current chapter by proposing how microgrids can help facilitate a pragmatic and dynamic approach to the global energy transformation, emphasizing efficiency, technological innovation, and adaptation to economic and social realities. We need a balanced plan that combines long-term goals with practical steps to achieve them.

Differentiating Sources of Production and Consumption: A Priority

The first step toward an effective energy transformation is to clearly distinguish between the challenges unique to various sources of energy production and consumption. Energy production must be optimized by type according to the specific needs of different end users, such as industrial and residential consumers. It is essential to apply new power generation, storage, and end-use electrification technologies where they can have the most significant, positive impact in terms of increasing energy independence while reducing emissions.

With power production, an increasing mix of sources is becoming available in addition to enhanced traditional ones. Within renewables, solar and hydroelectric are now established, traditional sources, while wind power, geothermal, and tidal are being developed with increasing speed. As for nuclear, while building a traditional nuclear power plant is prohibitively expensive in most use cases, small modular reactors and microreactors are being developed and offer that sector the possibility of a renaissance for large end users with stable power loads (e.g., data centers, steel and aluminum production, and oil refineries) and to eventually power microgrids as well. As for storage, while lithium ion battery packs could be now considered the traditional form of storage and is very well-suited to supporting loads that fluctuate greatly (e.g., a typical residence in North America), a veritable cornucopia of options is emerging, from sodium-ion batteries to flow batteries, thermal batteries, and iron-air batteries. Some of these technologies would be better suited for community smart grids or microgrids than traditional lithium-ion, due to safety concerns (lower risk of fire), longer life, and capacity. As these emerging technologies mature and proliferate, it will facilitate the optimization of electrical power production based on the type and size of the power demand at the lowest level of the demand ecosystem.

Regarding end use, many "low hanging fruit" exist for pairing with specific types of generation and storage to maximize self-sufficiency and minimize environmental impacts. Anything that is currently combusting fuel as its primary power source is a candidate, e.g., furnaces and other heaters, vehicles, stoves and ovens, drying machines, metallurgical production, concrete production and even datacenters (some have natural gas or diesel powered on-site generation). Each would have its own particular demand profile that is better suited for specific types of generation and storage. Also, the emergence of AI and its nearly insatiable demand for power that is forecast to outstrip available grid supply will force the deployment of larger systems previously considered to be "grid-scale" to serve just one facility or set of interconnected facilities. There may be opportunities for communities to partner in such private-sector deployments to greatly reduce the upfront costs they would face planning and installing a dedicated microgrid for themselves.

Maintaining and Improving Living Standards

People are worried that the energy transformation could affect living standards. A well-managed transformation can maintain and even improve living standards through more efficient technologies and lower long-term energy costs. It is important to encourage new ideas that reduce carbon footprints and improve quality of life. Smart home tech like thermostats, energy-efficient appliances, and home automation can make homes more comfortable and convenient while reducing energy use. Renewable energy solutions like solar panels and wind turbines can provide reliable and sustainable energy, reducing our reliance on fossil fuels and improving air quality. EVs reduce emissions and make driving quieter and smoother, which improves quality of life.

Green technologies must be available to all, including the most disadvantaged. Governments can help by making green technologies cheaper for low-income households. Community solar projects let people who cannot afford solar panels use renewable energy. Also, teaching people how to use green technology and incorporate it into their lives can help communities increase adoption rates and make it more appealing in general.

The energy transformation should focus on the sectors that can be transformed the quickest. Each sector must be addressed separately, considering its specific needs. Solutions for heavy industry are different from those for households. The industrial sector can reduce emissions and improve efficiency by using energy-efficient manufacturing processes, cleaner production technologies, and waste heat recovery systems. In homes, making buildings more energy-efficient, insulating existing homes better, electrifying gas-burning appliances, and using renewable energy can improve living standards, reduce emissions and reduce energy bills.

It is important to be flexible and change priorities when new technology and economics emerge. This means doing more research and developing new technologies. It also means making policies that can change with new technology and the market. Finally, it means working with the private sector to discover what new, affordable solutions are available to everyone.

Some people are worried about "going green" for a variety of reasons. To address these concerns about the energy transformation, we must provide information about the benefits and challenges of proposed changes, involve communities in decision-making, and establish support systems. By focusing on these aspects, we can manage the transformation in a way that maintains and improves living standards.

Transforming energy is one of the biggest challenges of the 21st century. But for it to be successful, it must be good for the environment and economical. It will not be sustainable if the end result is a reduction in the overall standard of living, while life is better for a minority of the population.

Conclusion: Empowering Prosumers and Shaping the Future

As we conclude Chapter 4, it is evident that the energy revolution is driven by the active participation of individuals and communities, transforming them into prosumers. This chapter has explored the multifaceted roles of prosumers in the evolving energy landscape, highlighting the impact of technological innovations, data management, modern economic models, and sustainability principles.

The concept of prosumers is central to the energy transformation, as individuals and communities take control of their energy production, consumption, and management. Technological advancements, such as smart grids, AI, and digital twins, are enabling more efficient and secure energy use, empowering prosumers to optimize their energy systems and contribute to a decentralized energy future.

Modern economic models and sustainability principles, including the circular economy and the reduction of carbon footprints, are reshaping the way we think about energy. By embracing these models, prosumers can create sustainable and economically viable energy solutions that benefit both the environment and the economy. The integration of renewable energy sources, energy storage systems, and smart grid technologies is essential for achieving energy autonomy and resilience.

International regulations on climate change and the SDGs provide a framework for supporting smart grids and decentralized energy systems. These regulations and goals emphasize the importance of collaboration, innovation, and sustainable practices in the global energy transformation. Operational strategies and change management are crucial for the successful transition to energy autonomy. By engaging communities, fostering continuous improvement, and adapting to changing circumstances, prosumers can build strong and resilient energy systems.

Future opportunities and business models for microgrids are emerging, offering new ways to encourage energy efficiency, create jobs, and develop innovative solutions. The exploration of gamification, new job roles, and business models such as MaaS and VPPs demonstrates the dynamic and forward-thinking nature of the energy landscape.

As we move forward, it is important to recognize that the energy revolution is not just about technology and economics; it is also about people and communities working together to create a sustainable future. By empowering prosumers and embracing innovative solutions, we can shape the future of energy and build resilient, self-sufficient systems that benefit everyone.

With a deeper understanding of the social, economic, and technological dimensions of the energy revolution, we are now ready to explore the practical applications of energy autonomy. Chapter 5, titled "The Greenberg Family's Journey Through Energy Autonomy", will provide a hypothetical case study of a family's drive for energy independence. We will follow the Greenberg family as they navigate the decision-making process, implement their energy solutions, and become part of an energy community. We will explore the challenges they face, the strategies they employ, and the benefits they achieve. The chapter will also discuss commercial applications and the future of energy innovation, highlighting how individual actions can contribute to the broader energy transformation.

References

1. https://www.ny.gov/sites/default/files/atoms/files/WhitePaperREVMarch2016.pdf.
2. https://www.cpuc.ca.gov/sgipinfo/.
3. https://www.brooklyn.energy/.
4. Gummesson, E. and Polese, F. (2009). B2B is not an island!. *Journal of Business and Industrial Marketing*, *24*(5/6), 337–350.
5. Fragnière, E., Sandoz, S., Abdenadher, N. and Moresino, F. (2024, May). Developing card-based surveys: Focus group insights for smart grid prosumer optimization. In: *2024 12th International Conference on Smart Grid* (*icSmartGrid*) (pp. 579–581). IEEE.
6. https://www.gov.uk/government/publications/transitioning-to-a-net-zero-energy-system-smart-systems-and-flexibility-plan-2021
7. https://www.smart-energy.com/industry-sectors/energy-grid-management/new-digital-twin-reference-library-to-further-decentralise-power-grids/.
8. Kumari, N., Sharma, A., Tran, B., Chilamkurti, N. and Alahakoon, D. (2023). A comprehensive review of digital twin technology for grid-connected microgrid systems: State of the art, potential, and challenges faced. *Energies*, *16*(14), 5525.
9. https://msites.epri.com/der-vpp-ferc2222/Information-Communication-and-Cyber-Security.
10. Naraindath, N.R., Kupolati, H.A., Bansal, R.C. and Naidoo, R.M. (2024). Data security and privacy, cyber-security enhancement, and systems recovery approaches for microgrid networks. In: *Modelling and Control Dynamics in Microgrid Systems with Renewable Energy Resources* (pp. 377–401). Academic Press.
11. https://blog.google/outreach-initiatives/sustainability/google-kairos-power-nuclear-energy-agreement/
12. https://www.popsci.com/technology/ai-more-energy/.
13. https://www.nbcchicago.com/news/business/money-report/why-big-tech-is-turning-to-nuclear-to-power-its-energy-intensive-ai-ambitions/.
14. https://www.iamsterdam.com/en/business/smart-mobility/amsterdam-is-a-top-european-city-for-shared-mobility.
https://www.citymonitor.ai/analysis/copenhagen-shared-zero-emission-mobility-european-cities/.

15. Ostrom, E. (1990). *Governing the Commons: The Evolution of Institutions for Collective Action*. Cambridge University Press.
16. Valencia, M., Bocken, N., Loaiza, C. and De Jaeger, S. (2023). The social contribution of the circular economy. *Journal of Cleaner Production*, *408*, 137082.
17. Laseter, T., Ovchinnikov, A. and Raz, G. (2010). Reduce, reuse, recycle, or rethink. *Strategy+ Business*, *61*, 1–5.
18. (a) What are scope 1, 2 and 3 carbon emissions? - National Grid Group.
    (b) Assumptions and References for Household Carbon Footprint Calculator ...
    (c) Household Carbon Footprint Calculator | US EPA.
    (d) Carbon calculator - GOV.UK.
    (e) What are Scope 1, 2 and 3 Carbon Emissions with Examples - Digit Insurance.
    (f) How to Calculate Your Carbon Footprint | PCMag.
    (g) CoolClimate Network Carbon Calculator.
    (h) Carbon Footprint Calculator | Climate Change | US EPA.
    (i) Carbon Footprint Calculator | Climate Change | US EPA.
19. https://www.holcim.com/sites/holcim/files/2024-04/28022024-holcim-climate-report-2023.pdf.
20. National Academies of Sciences, Engineering, and Medicine (2023). The role of net metering in the evolving electricity system.
21. National Academies of Sciences, Engineering, and Medicine (2023). The role of net metering in the evolving electricity system.
22. https://energy.ec.europa.eu/topics/energy-strategy/clean-energy-all-europeans-package_en.
23. https://eur-lex.europa.eu/legal-content/EN/TXT/?uri=uriserv:OJ.L_.2018.328.01.0082.01.ENG.
24. https://energy.ec.europa.eu/topics/energy-efficiency/energy-efficiency-targets-directive-and-rules/energy-efficiency-directive_en.
25. https://commission.europa.eu/strategy-and-policy/priorities-2019-2024/european-green-deal_en.
26. https://research-and-innovation.ec.europa.eu/funding/funding-opportunities/funding-programmes-and-open-calls/horizon-europe_en.
27. https://www.eia.gov/outlooks/aeo/.
28. https://www.energy.gov/policy/articles/department-energy-office-policys-2023-year-end-review-clean-energy-jobs-jump-tax.
29. https://www.energy.gov/sites/default/files/2023-05/DOE%20-%20100%25%20Clean%20 Electricity%20-%20Final.pdf.
30. https://www.whitehouse.gov/briefing-room/statements-releases/2023/05/10/fact-sheet-biden-harris-administration-outlines-priorities-for-building-americas-energy-infrastructure-faster-safer-and-cleaner/.
31. https://sdgs.un.org/goals.
32. United Nations. 2023. Sustainable Development Goals poster. United Nations. https://www.un.org/sustainabledevelopment/news/communications-material/.
33. https://smartgrid.ieee.org/bulletins/may-2023/overcoming-barriers-to-microgrid-development-a-review-of-policies-and-regulations.
34. Fragniere, E., Sandoz, S., Abdenadher, N., Moussa, M., Serugendo, G.D.M. and Glass, P. (2023, May). Improving the Social Acceptability of Microgrids. In: *2023 IEEE Renewable Energy and Sustainable E-Mobility Conference (RESEM)* (pp. 1–5). IEEE.
35. Vahidinasaheb, V. and Mohammadi-Ivatloo, B. (2023). *Energy Systems Transformation: Digitalization, Decarbonization, Decentralization and Democratization*. Springer Nature.
    Asif, M. (Ed.). (2022). *The 4Ds of Energy transformation: Decarbonization, Decentralization, Decreasing Use, and Digitalization*. John Wiley & Sons.
36. *Source:* Ngo, J., Fragnière, E., Larpin, B. and Sahut, J.M. (2023, March). Gamification to Stimulate Green Behaviors in Cities. In: International Conference on Civil Engineering (pp. 195–205). Singapore: Springer Nature Singapore.

37. Karimov, A., Durst, S., Saarela, M., Gerstlberger, W. and Wahl, M.F. (2024). A two-phase systematic literature review on the use of serious games for sustainable environmental education. *Interactive Learning Environments*, 1–22. https://doi.org/10.1080/10494820.2024.2414429.
38. Zhang, S., Liu, Y. and Chen, Y. (2024). Enhancing ESG learning outcomes through gamification: An experimental study. *Sustainability*, *16*(10), 4066. https://doi.org/10.3390/su16104066.
39. Ngo, J., Fragnière, E., Larpin, B. and Sahut, J.M. (2023, March). Gamification to Stimulate Green Behaviors in Cities. In: *International Conference on Civil Engineering* (pp. 195–205). Singapore: Springer Nature Singapore.
40. Ngo, J., Fragnière, E., Larpin, B. and Sahut, J.M. (2023, March). Gamification to stimulate green behaviors in cities. In: *International Conference on Civil Engineering* (pp. 195–205). Singapore: Springer Nature Singapore.
41. https://www.chaostheorygames.com/work/energy-transformers#:~:text=Energy%20Transformers%20is%20an%20immersive,an%20exhilarating%20journey%20across%20Australia.
42. https://www.georgiapower.com/company/news-hub/company-news/georgia-power-launches-new-mobile-app-for-customers.html.
43. https://thinkenergy.org/consumers/families/interactive-games/.
44. https://energy.techno-science.ca/en/energy-games.php.
45. How many jobs could the clean energy transformation create? and The renewable energy transformation is creating a green jobs boom.
https://microgridnews.com/reports-show-microgrids-create-jobs-and-economic-growth/.
46. FACT SHEET: The Bipartisan Infrastructure Deal Boosts Clean Energy Jobs ...
FACT SHEET: President Biden's Budget Lowers Energy Costs, Combats the ...
Biden-Harris Administration Announces $72 ... - Department of Energy
Poll: Biden touts his four major infrastructure and clean energy laws. Voters doubt they're working.
47. https://www.energy.gov/sites/default/files/2022-12/Topic%207%20Report.pdf.
48. Shahzad, S., Abbasi, M.A., Ali, H., Iqbal, M., Munir, R. and Kilic, H. (2023). Possibilities, challenges, and future opportunities of microgrids: A review. *Sustainability*, *15*(8), 6366. https://www.mdpi.com/2071-1050/15/8/6366.
49. https://www.energy.gov/sites/default/files/2022-12/Topic%207%20Report.pdf.
50. https://www.startus-insights.com/innovators-guide/microgrid-trends/.
51. https://www.technologyreview.com/2024/02/07/1087836/how-virtual-power-plants-are-shaping-tomorrows-energy-system/.
52. https://www.eia.gov/todayinenergy/detail.php?id=48296.
53. Le-Anh, T., Nguyen, M.D., Nguyen, T.T. and Duong, K.T. (2023). Energy saving intention and behavior under behavioral reasoning perspectives. *Energy Efficiency*, *16*(2), 8. https://link.springer.com/article/10.1007/s12053-023-10092-x.
http://www.cres.gr/behave/framework_theory_2.htm
54. Social Sci LibreTexts. (n.d.). Changing attitudes by changing behavior. In: *Principles of Social Psychology*. Retrieved from https://socialsci.libretexts.org/Courses/Lumen_Learning/Book%3A_Principles_of_Social_Psychology_%28Lumen%29/05%3A_4._Attitudes_Behavior_and_Persuasion/05.4%3A_Changing_Attitudes_by_Changing_Behavi.
55. Brehm, P.A., Johnston, S. and Milton, R. (2024). *Backup Power: Public Implications of Private Substitutes for Electric Grid Reliability*. The University of Chicago Press.
56. Hirsch, A., Parag, Y. and Guerrero, J. (2018). Microgrids: A review of technologies, key drivers, and outstanding issues. *Renewable and Sustainable Energy Reviews*, *90*, 402–411. https://doi.org/10.1016/j.rser.2018.03.040.

5 Greenberg Family's Journey through Energy Autonomy

This narrative-driven chapter makes the theoretical personal through the fictional yet instructive journey of the Greenberg family. The family decides to transition their suburban home to a microgrid. Their story illustrates the practical challenges, decision points, and triumphs that accompany the pursuit of energy autonomy. Readers follow the Greenbergs step by step, from their initial frustration with rising energy bills and unreliable grid performance to the final commissioning of their customized microgrid. The chapter covers technical design, cost analysis, contractor selection, and navigating utility interconnection. Earlier concepts, such as load forecasting, solar-battery integration, and demand response, are brought to life in a relatable setting. Family dynamics are woven into the technical narrative as different family members champion specific goals, such as cost savings, reducing their carbon footprint, and increasing resilience during outages. Unexpected challenges, such as zoning restrictions, equipment delays, and financing hurdles, are transparently addressed, offering real-world lessons for readers. Over time, the Greenbergs evolve from passive consumers to active prosumers who participate in energy trading and educate their neighbors about local energy solutions. Their journey is a microcosm of the broader movement toward decentralized, people-powered grids. The chapter concludes with reflections on how microgrid adoption can transform infrastructure, individual mindsets, and community relationships, reinforcing the book's central message that energy autonomy is achievable and empowering.

In this chapter, we explore hypothetical residential and commercial/industrial examples of energy autonomy. The residential example depicts various scenarios, including rural or urban settings, a single household, an entire energy community district, or even a vacation destination. In the commercial/industrial example, different settings such as malls or factories are considered. The aim is to cover a wide range of possible situations and applications regarding energy autonomy. This chapter is intended to help the reader relate to what the authors have presented so far to everyday situations.

The Greenberg family's journey towards energy independence is a practical example of the ideas discussed in the previous chapters. Their story will show how microgrids can provide energy security, reduce carbon footprints, and offer financial savings. Readers will learn how to apply these principles to their own lives by following the Greenbergs' journey. This chapter will also highlight the importance of community engagement and the role of technological innovation in achieving energy independence.

DOI: 10.1201/9781032689845-5

Introduction: A Family's Drive for Energy Independence

Let's give a warm welcome to the Greenberg family! The parents are active professionals in the fields of IT and supply chain management and have two children aged 18 and 21. They are accustomed to teleworking a few days a week from home, especially after the COVID-19 pandemic. Additionally, they are environmentally conscious and want to reduce their energy consumption to minimize CO_2 emissions. They have been informed of a new development managed by a homeowners association (HOA) in the nearby countryside that will be powered by a microgrid using solar power generation. The financing terms are favorable, but the purchase comes with atypical terms in the HOA bylaws and rules related to the microgrid being installed as part of the development. The family has reviewed the new terms and is considering whether they are acceptable for a residence they plan to inhabit for an indefinite period.

They scheduled a meeting with the manager at a completed sample home. The manager explains that the house is equipped with a smart meter that will track electricity consumption real time. The HOA includes rules related to electricity consumption in addition to other typical communal rules such as noise, exterior decor, trash, parking, etc. The solar panel roofing system in this sunny location will provide an average of half the needed electricity in the winter. During the summer months, there is a higher demand for electricity due to the need for air conditioning. As a result, there is a greater dependence on the grid during peak periods when many people are consuming electricity simultaneously.

Consequently, residents may be required to reduce their electricity consumption by postponing tasks such as running the dishwasher or laundry machines/dryers, or by setting the air conditioning to a higher temperature. While the manager assures the Greenburgs that electricity bills will be much lower compared to typical homes, the Greenberg family must accept potential changes to their electricity consumption habits. Their role in the economy may significantly change. Previously, they were purely consumers of electricity. Now, they would become prosumers.

Overall, the Greenbergs must reduce electricity usage, especially at peak times, for the system to work optimally. Everyone in the community must follow these rules to get the most value out of the microgrid. This sense of community commitment would help people in the community get to know each other better and work together toward a common purpose. The HOA would manage the microgrid, ensuring energy usage rules are followed. During outages or high demand, energy quotas may be used to limit energy use. This can be challenging, because the whole community has to work together to manage their energy use well in order to make sure everyone has what they need [1].

However, the benefits of living in a community with a well-managed microgrid are significant. The Greenbergs choose to move to another community where they pay a slightly higher premium for their home but have a system with storage that can offset peak demand, guaranteed energy savings, and dividends when supply exceeds demand. They had initially thought about a different community that offered solar generation but lacked integrated storage. After thinking about it, they decided not to go with that first choice. The second community had advanced storage, which saved energy and used it during times of high demand. This was an important factor. This made it more reliable, saved money in the long term, and matched their aims for energy resilience and independence. In this community, the microgrid is equipped with advanced energy storage solutions to ensure that excess energy produced during peak periods is stored and used during periods of high demand. This not only ensures a stable and reliable power supply but also results in significant cost savings on electricity bills. The community's commitment to renewable energy makes it environmentally friendly, reducing its carbon footprint and promoting sustainability. In addition, the community has a well-structured financial model that provides dividends to residents when energy supply exceeds demand. This allows the Greenbergs to financially benefit from their investment in the community's energy infrastructure. The collaborative approach to energy management fosters a sense of unity and shared responsibility, making the community a supportive and engaging place to live.

Overall, the Greenbergs find that the slightly higher home purchase premium is well worth the benefits of living in a community with a guaranteed energy savings system, reliable energy storage, and the potential for financial dividends. This makes their decision to move to this community an easy and attractive choice. After careful consideration, the Greenbergs decided to move into the community with the advanced microgrid system. They were impressed with the guaranteed energy savings, reliable energy storage, and the potential for financial dividends. The community's collaborative approach to energy management and commitment to sustainability further solidified their decision. The slightly higher premium they paid for their home was well worth the benefits of having a stable and cost-effective energy solution. The Greenbergs are pleased to be part of a forward-thinking community that values environmental responsibility and energy independence.

Now let us turn back the clock by a few months. In fact, the Greenberg family had already started down the road to becoming prosumers. They just had not realized it yet!

The Greenberg Family's Energy Adventure!

The family was excited to start their journey. They were packing their camping gear into their SUV and trailer. Emma, the daughter, asked her father if he had packed the solar panels. Mr. Greenberg smiled as he patted the sleek, portable 1 KWh power station nestled in the corner of the trailer. He explained that the power station would provide all the energy they needed, straight from the sun. The Greenbergs were embarking on an energy adventure, seeking self-sufficiency in a secluded spot nestled deep within nature, far from the hustle and bustle of city life. However, this was not just any vacation. It was a test of their recently acquired knowledge. Only a week prior, they had completed a crash course in "microgrid 101", learning the fundamentals of energy autonomy. Now, equipped with their portable power station and a newfound understanding of energy management, they were prepared to apply theory to practice.

The journey was fun. They stopped to admire the views and stretch their legs. At each stop, they talked about their plans and made sure they were ready. When they arrived at their campsite, the family set up camp together. They connected the solar panels to the power bank. The panels absorbed the sunlight, provided power to their devices and charged the power bank's battery. The power bank stored excess electricity generated during the day for use at night or during cloudy periods. The solar panels were placed in a sunny spot with minimal shading and angled to get the most sunlight. The family used an energy management system (EMS) integrated into their power bank to monitor their energy use thereby helping them to track and save energy. The EMS ≈. They charged their devices during the day and used energy-efficient lights at night. The power bank gave them a backup power source in case they had an unplanned need for extra power. If it was cloudy, the stored energy meant that essential devices like lights, a small refrigerator, and communication devices stayed on.

The Greenbergs used solar power to charge their phones and power a laptop. They also used solar power to cook, which was safer, cleaner, and more convenient than using a campfire or propane or charcoal grill for cooking. They saved energy by turning off lights, using natural light, and not using appliances that use a lot of energy when the sun was not shining. The adventure was a learning experience for the whole family. They learned about solar panels, managing energy use, and living sustainably. Emma and her brother learned about renewable energy and how using it responsibly can help the environment.

In the evening, they gathered around a campfire. The fire and soft-white, highly-efficient LED string lighting made the place feel cozy. Mr Greenberg looked at his family with pride. "This is it," he said, "living off the grid with solar power." "We're pioneers of the modern age." The family roasted marshmallows and enjoyed the simple, peaceful, serene surroundings that lacked the background roar of a gasoline fueled generator that would have normally been required to power the electric equipment the Greenbergs were using.

As they went to sleep, the Greenbergs felt happy. They had learned a lot and were living proof that sustainable living was not only possible but also fun. This was just the beginning of their journey towards energy independence and a more sustainable future.

The Epiphany

The Greenbergs started on their journey to energy autonomy with curiosity and a growing awareness of their impact on the environment and energy costs. That summer, the Greenbergs got a shockingly high electricity bill. The air conditioning used during the heatwave of unprecedented duration and extremity increased their energy consumption considerably above average. The family realized they needed a more sustainable energy solution because of an increasing number of grid outages and the increasingly frequent and urgent news reports about the deleterious effects of climate change. They also often needed to charge their EV during peak periods. This exemplified their need for a better, more reliable energy system [2].

The first step was education. Mr. Greenberg and his wife spent evenings researching renewable energy options. They soon discovered that microgrids—localized energy systems capable of generating and storing energy independently—offered both environmental and financial benefits. What appealed most to them was the promise of energy resilience, particularly during the increasingly frequent grid outages their suburban neighborhood experienced due to storms and an aging infrastructure.

Decision-making Process

Making the decision to adopt a microgrid was not instantaneous. The Greenbergs were cautious and took several important steps before fully committing to their new energy system.

1. Identifying Needs and Setting Goals

Before investing in a microgrid, the family had to first identify their energy consumption patterns. They used data from their smart meter to track how much energy they were using daily, particularly during peak times. They also calculated their monthly energy costs to evaluate the potential savings a microgrid could offer.

Mrs. Greenberg was particularly focused on energy sustainability and wanted to reduce the household's reliance on fossil fuels. Meanwhile, Mr. Greenberg prioritized energy independence and resilience, hoping that a microgrid would provide backup power during frequent outages. Together, they established three key goals:

- **Reduce Reliance on the Traditional Power Grid:** Reducing reliance on the traditional power grid involves generating and managing your own electricity through renewable energy sources like solar panels or wind turbines. This shift not only decreases dependency on external power providers but also mitigates the impact of grid outages and fluctuations. By producing their own energy, households and communities can achieve greater energy independence, ensuring a more stable and predictable supply of electricity tailored to their specific needs.
- **Lower their Carbon Footprint and Contribute to Environmental Sustainability:** Lowering the carbon footprint means reducing the amount of greenhouse gasses, particularly CO_2, that are emitted into the atmosphere as a result of energy consumption. By adopting renewable energy sources such as solar or wind power, which produce no emissions, individuals and communities can significantly decrease their environmental impact. This transition supports broader efforts to combat climate change, preserve natural ecosystems, and promote a healthier planet for future generations.

- **Ensure Energy Security during Outages or Emergencies:** Ensuring energy security during outages or emergencies involves having a reliable backup power system that can maintain essential services when the main grid fails. This can be achieved through energy storage solutions like batteries that store excess energy generated by renewable sources. During power outages caused by storms, natural disasters, or other disruptions, these systems can provide continuous power to critical appliances and systems, enhancing resilience and safety for households and communities.

2. *Initial Research*

Armed with their goals, the Greenbergs dove into researching various energy systems. They explored options like rooftop solar panels, energy storage systems (ESS). They read articles, watched documentaries, and joined online forums for prosumers—households that both produce and consume energy. The family quickly realized that a solar-powered microgrid with battery storage suited their needs best.

They also reviewed case studies of other families who had adopted microgrids, which helped them understand the financial and practical implications of energy autonomy. This research led them to the realization that while the upfront cost was significant, the long-term savings and environmental benefits outweighed the initial investment.

3. *Consulting Experts*

The Greenbergs knew their research had limits. Before meeting with the microgrid developer, they asked experts to assess the pros and cons of adopting a microgrid. They met with solar and microgrid experts, who explained the different types of systems. The experts explained the pros and cons of different microgrid setups and helped the Greenbergs understand the technical and financial aspects of going green.

During their consultations, they learned about key components like:

Solar Panels: The consultants recommended installing high-efficiency monocrystalline panels for maximum energy generation. The number of high-efficiency monocrystalline panels you need depends on several factors:

- *Energy Consumption:* Determine your household's daily energy usage in kilowatt-hours (kWh). For example, if your home uses 30 kWh per day, you'll need enough panels to generate this amount.
- *Panel Efficiency:* High-efficiency monocrystalline panels typically have an efficiency of around 17–24%. This means they can convert 17–24% of the sunlight they receive into electricity.
- *Sunlight Availability:* Consider the average sunlight hours your location receives per day. For instance, if you get 5 hours of peak sunlight daily, a 300-watt panel would generate about 1.5 kWh per day (300 watts × 5 hours).

- *System Size:* To generate 30 kWh per day with 300-watt panels, you would need around 20 panels (30 kWh/1.5 kWh per panel).
- *Battery Storage:* To store excess energy for night-time use, the engineers suggested a 10 kWh lithium-ion battery system.
- *Average Night-Time Consumption:* Calculate your household's energy usage during the night. For example, if your home uses around 1–2 kWh per hour during the night, a 10 kWh battery could last 5–10 hours.
- *Energy Efficiency:* Ensure your home is energy-efficient. Using energy-efficient appliances and LED lighting can help extend the battery's capacity.
- *Backup Needs:* If you need the battery for backup during power outages, consider how long you typically need backup power.
- *Solar Panel Output:* If you have solar panels, the battery can store excess energy generated during the day for use at night, enhancing overall efficiency.
- *Smart Energy Management:* The family learned about EMS that would automatically monitor and optimize their energy consumption.

The smart meter shows how much energy is used and generated in real time. It tells the EMS when and how much energy is being used. This helps the family adjust their usage patterns, such as running appliances during off-peak hours to further reduce costs. The combination of the EMS and the smart meter provides cost savings, energy efficiency, and convenience through automated control and real-time monitoring. The EMS could also schedule appliances to run during off-peak hours to save costs. This setup was energy-efficient, cost-effective, and reliable.

The experts also performed a detailed energy audit of the Greenberg home, measuring factors like roof space, shading, and daily sun exposure to ensure that their home was suitable for solar energy. The Greenbergs had enough roof space for 28 high-efficiency panels. Each panel needs about 17.5 square feet. They installed 20 panels, with room for more.

The audit showed that there was little shading, with only a few trees casting shadows in the morning and afternoon, so the house gets about 5 hours of peak sunlight a day. The 20 panels would produce about 30 kWh a day, matching the Greenbergs' average energy use.

The battery system stored excess energy for night-time use. The Greenbergs saved on electricity costs by using solar energy and the EMS. If peak electricity rates were $0.30 per kWh, using 10 kWh from the battery during peak hours saved them $3.

4. Evaluating Costs and Incentives

One of the primary concerns of the Greenbergs was the financial implications of installing a microgrid. However, after consultations, they discovered that various financial incentives could help offset the initial investment. The family took advantage of:

- Federal solar ITC, which offered up to 30% off installation costs.
- Local utility rebate programs that provided subsidies for battery storage systems.
- State-level net metering and REC markets, which allowed them to sell excess solar power back to the grid and obtain revenue from REC sales for producing 100% renewable energy, providing long-term financial benefits.

Long-Term Considerations

Before making their final decision, the Greenbergs also considered the long-term benefits and responsibilities associated with managing a microgrid. They asked key questions:

- How much maintenance would be required for the solar panels and battery system?
- How quickly would they recoup their initial investment in terms of energy savings?
- What lifestyle changes would be necessary to maximize their energy independence?

The Greenbergs thought about the long-term benefits and responsibilities of managing a microgrid before making their final decision. They asked key questions to help them make an informed choice.They looked into the maintenance needed for the solar panels and battery system. Solar and battery systems usually need maintenance three times over 20 years. This includes checking the panels and battery system. The battery and inverter usually need replacing after 10–12 years.

Next, they calculated how quickly they would save money. They divided the initial cost by the annual savings. If the initial cost was $18,000 and they saved $2,100 a year on electricity, it would take about 8.5 years to recoup the cost. After 8.5 years, they would start saving and accruing money.

They also thought about how to make the most of their new energy independence. To get the most out of their solar and battery system, they needed to change how they used energy. This meant using dishwashers and washing machines during off-peak hours and being careful about energy use. The smart meter in their home helped them make these adjustments.

Through these discussions, they realized that adopting a microgrid was not just about installing new technology—it was about rethinking how they used and managed energy in their everyday lives.

Factors Influencing the Greenberg Family's Decision to Adopt a Microgrid

1. **Rising Energy Costs:** With energy bills steadily increasing, the family saw a microgrid as a way to control their costs by producing their own power, and eventually eliminating electricity bills through net metering.
2. **Grid Reliability and Resilience:** Frequent outages, particularly during extreme weather events, pushed the family to seek a more reliable, resilient

solution. A microgrid could ensure their home remained powered during grid failures.

3. **Environmental Concerns:** Both Mr. and Mrs. Greenberg were motivated by a desire to reduce their carbon footprint. The environmental impact of burning fossil fuels was an increasing concern for them, and they saw renewable energy as a way to make a tangible difference.
4. **Government Incentives:** The availability of tax credits and rebates played a significant role in making the project financially viable. They realized that this was an optimal time to make the transition while incentives were still available.
5. **Energy Independence:** The appeal of being energy independent—generating, storing, and managing their own electricity—was a major factor. They viewed microgrid technology as a way to take control of their energy future.

The Greenbergs' decision-making process shows how important it is to do thorough research, consult experts, and plan carefully. By understanding their energy needs, exploring renewable options, and navigating regulatory requirements, they were able to make informed decisions that suited their goals of sustainability and cost-effectiveness. This shows other people that they too can make the same choices.

The Greenberg Home as Part of an Energy Community

The Greenberg Residence is part of a community powered by renewable energy. All neighbors have invested in energy storage systems. In the event of a grid outage or emergency, energy reserves ensure that essential appliances and systems can continue to function, providing security and peace of mind to residents. Additionally, a microgrid within the community detects grid disturbances and seamlessly transforms to island mode, isolating the community from the broader grid to prevent potential safety hazards, such as voltage fluctuations or power surges. This ensures that residents are prepared for unforeseen disruptions and emergencies, such as natural disasters or grid failures. The islanding capability of the Greenberg Community enhances the reliability of the community's energy supply and contributes to its safety and resilience. The Greenberg Residence exemplifies environmental sustainability, safety, and resilience in modern communities. By embracing islanding technologies and fostering a culture of preparedness, residents have created a haven where safety, sustainability, and community thrive together.

EVs as Energy Storage: The Greenberg Experiment

The Greenberg family recently accepted the invitation to participate in a groundbreaking research project led by a local university. The project involved using EVs as storage devices. Despite initial skepticism from the Greenbergs and their neighbors, the family became confident in their ability to contribute to the project's success. The university research team provided informative sessions and practical demonstrations to help the audience understand how EVs can contribute to energy storage and grid stability. They explained that EV batteries can be used as mobile energy storage units, charging during off-peak hours when renewable energy generation is abundant and discharging when demand on the grid is high. Technological integration was crucial in making this vision a reality. The research team collaborated with EV manufacturers and energy community stakeholders to develop software solutions that seamlessly integrated EVs into existing microgrids. This enabled bidirectional energy flow between vehicles and the grid, allowing EV owners like the Greenbergs to contribute surplus energy back to the community when needed.

The Greenbergs acquired an EV and installed a bidirectional charging station at their residence. They reduced their carbon footprint through sustainable transportation and participated in grid-balancing initiatives. The university research project conducted field trials, collecting valuable data on EV charging patterns and energy storage performance, which indicated promising outcomes. The integration of EVs as storage devices would optimize energy management within the community. This would reduce peak demand periods and enhance grid stability. In conclusion, the Greenbergs' EV journey demonstrates the transformative potential of EVs beyond transportation. Collaboration among residents, academia, and industry stakeholders can transform EVs from mere vehicles into valuable assets in decentralized energy systems. This could pave the way for a cleaner, more sustainable energy future while enhancing lives by providing more people with energy resilience and autonomy.

The Mall's Microgrid Solution

The Greenbergs are at their favorite mall, where innovation and sustainability meet. The mall has a new microgrid system with solar panels, batteries, and energy controls. This system makes sure there is always enough power and manages how it is used. A storm damages the power lines while the Greenbergs are shopping. The mall's microgrid keeps stores and restaurants running smoothly by taking over, managing the energy flow and keeping shoppers comfortable. The Greenbergs do not notice any change in their shopping experience and are unaware of the technology at work behind the scenes.

When the Greenbergs notice a local power outage notification on their phones, they pause to inquire with a mall manager about the uninterrupted power and listen with great interest as the manager explains the resilience and reliability of the microgrid system. The manager highlights the benefits of the microgrid, emphasizing its role in ensuring business continuity, safety, and reliability. By relying on the microgrid, the mall provides a safe environment for shoppers and businesses even in adverse conditions. In addition, the integration of solar panels and energy storage into the microgrid and total property electrification eliminates reliance on fossil fuels and lowers the mall's carbon footprint, which appeals to the Greenbergs' values of sustainability.

When the Greenberg family completes their shopping trip, they leave with a newfound appreciation for the mall's microgrid solution. Beyond convenience and sustainability, they now recognize the critical role microgrids play in ensuring security, reliability, and business continuity in today's unpredictable environment. Inspired by their experience, they are eagerly exploring ways to integrate microgrid solutions into their own community, envisioning a future where resilience and sustainability go hand-in-hand.

A Few Months Later at the Greenberg's

At the Greenberg Residence, the Greenberg family eagerly awaits their first utility bills after moving into their new home. Having been briefed on the benefits of smart meter usage and demand-side management, they were hopeful but uncertain about the potential savings compared to their previous residence. Upon receiving their bills, the Greenbergs are pleasantly surprised by the significant savings they had achieved. Compared to their previous living situation, their energy expenses notably decreased, sparking curiosity about the factors contributing to this positive outcome.

As the annual HOA meeting approached, anticipation filled the air. Residents gathered to discuss the results and gain insights into the community's collective energy performance. The manager, adeptly addressing the multitude of questions, elucidated the key drivers behind the impressive savings. The manager emphasized the importance of variable elements, such as weather patterns and individual consumption behaviors, in influencing energy costs. While some factors were beyond their control, the implementation of demand-side management strategies, guided by the smart meter data, played a pivotal role in optimizing energy usage and minimizing expenses.

Despite acknowledging the potential variability from year to year, the manager underscored the long-term commitment required from the energy community to sustain these positive outcomes. Through continued collaboration and shared responsibility, the community could navigate challenges and maintain its trajectory towards energy efficiency.

As the meeting drew to a close, a sense of pride permeated the room as the manager unveiled the grand finale: the dramatic reduction in the community's CO_2 footprint. Through their collective efforts, the residents had not only achieved financial savings but had also made a tangible impact on environmental sustainability, reinforcing their commitment to a greener future.

In the end, Greenberg's neighborhood stood as a shining example of the transformative power of energy efficiency and community collaboration. Through education, innovation, and collective action, the residents had not only improved their quality of life but had also paved the way for a more sustainable and prosperous future for generations to come.

The Greenberg family has always been passionate about sustainability and energy independence. Their journey toward energy autonomy began with a high electricity bill and frequent power outages, which led them to explore renewable energy options. After extensive research and consultation, they decided to move to a community with an existing microgrid so they could benefit from generating, storing, and managing their own power as part of a shared system. This decision provided them with energy security, reduced their carbon footprint, and offered financial savings. Meanwhile, their friends, the Johnsons, faced similar challenges and decided to take a different approach. The Johnson family moved to a rural village of 1,500 people where the entire community had come together to build a solar panel farm to provide electricity. This collective effort not only provided a stable and sustainable energy supply for the village but also fostered a sense of community and shared responsibility.

For the Johnsons, this decision was made collectively, reflecting the village's strong commitment to sustainability and shared values. In this village, solar panels were installed on existing roofs and over parking lots to conserve space and minimize disruption. The electricity generated by these panels was owned by the community, fostering a sense of shared responsibility and benefit. A professional company was hired to manage the solar panels and storage systems, ensuring efficient operation and maintenance. Importantly, the project was financed through a combination of government grants, community investment, and loans. This approach meant that residents did not have to pay any upfront costs, making it easier for everyone to participate. The system operated like a utility, with residents paying for electricity at a reduced rate compared to traditional sources. However, while the electricity was cheaper than before, there was a premium for the added resiliency and autonomy the system provided. This arrangement helped manage electricity use and generation, improving reliability and reducing dependence on external electricity sources. Although the village still used grid electricity, the solar farm significantly lowered electricity costs over time and provided a more sustainable energy solution.

Ultimately, both paths offered significant benefits. The Greenbergs' decision to move to a community with an existing microgrid gave them the benefits of direct control over their energy use and a personal sense of accomplishment. They experienced lower electricity bills and a reduced carbon footprint, similar to their previous home microgrid. However, the sense of community and shared responsibility was much stronger in the Johsnons' new village. Their community's decision to build the solar farm demonstrated strong local support and alignment with their values and needs. Unlike the Greenbergs's community, where the HOA and a professional service company managed their own microgrid, living in the Johnsons' village meant being part of a larger, coordinated effort. Both approaches show how communities can use solar power supplied microgrids to save money, help the environment, and become more self-sufficient.

Empowering the Next Generation in the Energy Transformation

Kevin and Jennifer, the two college students of the Greenberg family, find themselves at a job fair dedicated to careers in the energy sector. Excitement fills the air as they explore the various stands, each offering insights into the diverse array of professions and opportunities emerging in the evolving energy landscape.

At every stand they visit, Kevin and Jennifer are greeted with enthusiasm and passion. Representatives eagerly share stories of innovation, growth, and impact, igniting a spark of curiosity and ambition within the young visitors. From renewable energy technicians to sustainable system engineers, the possibilities seem endless, and Kevin and Jennifer eagerly jot down their preferred career choices.

As they navigate through the fair, Kevin and Jennifer are acutely aware of the significance of their choices. The energy transformation not only promises economic prosperity but also holds the key to addressing pressing environmental challenges and building a more sustainable future for generations to come.

Their parents, observing from a distance, swell with pride and optimism as they witness their children's enthusiasm for the opportunities that lie ahead. In this moment, they recognize that the future is indeed bright, not just for their family but for the entire new generation.

As the job fair draws to a close, Kevin and Jennifer leave with a renewed sense of purpose and determination. Armed with a vision for a sustainable future and a passion for making a difference, they embark on their journey, ready to play their part in shaping a world powered by innovation, sustainability, and opportunity.

Crisis Situation and the Importance of Business Continuity Planning

In 2024, a storm named Helene, a powerful hurricane, caused widespread damage in the southeastern US. The storm made landfall in Florida's Big Bend region with winds of 140 mph, uprooting trees, submerging neighborhoods, and cutting electricity to over 4 million homes and businesses. The aftermath of Helen showed the need for energy resilience and robust business continuity plans to cope with disasters.

After the storm, communities were in the dark. Essential services like lighting, air conditioning, and heating were no longer working. The elderly and vulnerable struggled to cope without support. The storm also caused enormous economic damage to businesses and communities. This shows why it is important to have good business continuity planning (BCP) to make sure there is enough electricity for basic needs during a blackout.

BCP is a protocol that enables the energy community to rely on its own production and storage capacities to guarantee a minimum availability of electricity for basic needs in case of a blackout. The Business Continuity Plan (BCP) This ensures that essential services, such as lighting, air conditioning for the elderly, and heating during winter, are available and people remain comfortable and safe. Microgrids are excellent components of a BCP and address the critical aspects of how to maintain a power supply for days or weeks in the event of a grid outage. So critical are these that the United States Department of Defense has directed all military installations by 2030 to ensure energy resilience and have the ability to island their power supply for a minimum of two weeks [3].

A prepared community is more likely to recover quickly [4]. However, an effective BCP is not static. It requires regular updates and testing. Energy communities should hold regular drills, black start exercises and simulations to make sure everyone knows the plan and that it works in different situations. Continuous improvement based on lessons learned is crucial for maintaining the plan's relevance and effectiveness. To keep the plan relevant and effective, we must learn from our mistakes and practice regularly. Continuous improvement is important for keeping the plan relevant and effective.

A Business Impact Assessment (BIA) is also important. A BIA helps identify how a disruption could affect business operations and which functions are most important to restore. This assessment helps create BCPs by showing where there are weaknesses and how to use resources. A BIA would have identified the most critical services during Helene, such as healthcare facilities and emergency response units.

Commercial Applications

Mrs. Greenberg was impressed by the rows of solar panels stretching as far as the eye could see. It was her first day at work in the newly implemented solar panel factory in the heartland of the US, and she felt excited. It is a symbol of progress and sustainability in the heart of the nation. The factory is located in a region with abundant sunlight and is poised to harness nature's bounty and transform it into clean, renewable energy. Upon entering the industrial space, Mrs. Greenberg was met with the sound of machinery and the rhythmic clang of metal against metal echoing through the halls. The rooftops were adorned with massive solar arrays that captured the sun's rays and converted them into electricity to power the production process. The rooftop panels were not just for display, but were an essential component of the factory's microgrid system, providing the energy required to maintain smooth operations. The production floor was equipped with state-of-the-art machinery and staffed by skilled workers, such as Mrs. Greenberg, who were committed to the goal of creating a more sustainable future. From raw materials to finished products, the manufacturing process was carefully orchestrated to minimize waste and maximize efficiency.

The factory's commitment to sustainability was evident in the solar panels that adorned every available surface exposed to sunlight, providing shade and generating clean energy to offset the facility's power needs. Charging stations for EVs were present in the parking lot, providing free electricity to employees who had switched to eco-friendly transportation. Mrs. Greenberg felt a sense of purpose as she joined her colleagues on the production line, surrounded by cutting-edge technology and driven by a shared vision of a better tomorrow. As the day progressed, Mrs. Greenberg worked with renewed vigor, proud to be part of an industry that is shaping the future of energy and making a tangible difference in the world.

The factory's energy independence affected not only its own operations but also the wider economy. The factory used solar panels to make its own electricity, so it did not need to use the traditional power grid. This meant the company had a reliable energy supply, protected from price changes and outages. The company saved a lot on energy bills and used the money to invest in new technology and training. The factory's commitment to sustainability attracted investors and customers who care about the environment. Using renewable energy in production also helped reduce the factory's carbon footprint. This helped the company to be part of the effort to mitigate the effects of climate change and also to find new customers in markets that care about the environment.

The factory's economic benefits spread beyond itself. The local community saw more jobs, both at the factory and in the supply chain. Schools started training people in renewable energy to help the green economy. The factory's energy independence showed other businesses that investing in renewable energy is possible and beneficial. As more companies did the same, the economy became more resilient and sustainable. Moving to decentralized energy systems eased pressure on the utility grid, making energy more secure.

In short, Mrs. Greenberg's company shows how energy independence can help the economy, environment, and society. By using renewable energy, the factory not only made itself more efficient, but also helped create a sustainable future for the industry and the economy.

Small and Medium Enterprises (SMEs)

Other contexts include commercial and agricultural operations, essentially anywhere there is space for a large-scale plant to be installed for self-consumption of electricity and water heating, as well as to sell excess electricity output. The Greenberg family, always on the lookout for sustainable investments, decided to buy stocks in an innovative company. The manufacturing plant has a large roof surface area that accommodates an electrical plant with a significant amount of KW. This feature is particularly relevant for remote companies that may experience unreliable electricity, such as this manufacturer of high-end furniture in the middle of a forest in Montana. Being near the raw materials and operating on a short cycle is ideal for the company, while its remoteness makes reliable grid power supply a challenge. The company is finally attempting to achieve maximum independence from the grid. However, there will remain an interconnection agreement with the grid that allows the company to purchase extra electricity during peak demand and sell excess electricity back to the grid. The company plans to make significant investments in energy storage with batteries, and all employees would be encouraged to purchase electric vehicles to receive free electricity. Additionally, the company would utilize massive heat pumps for heating purposes.

As investors, the Greenbergs were keen to understand the financial implications and potential returns. They reviewed the following estimates for Carpentry Furniture: Solar Power System: $500,000 to $750,000; Energy Storage: $400,000 to $600,000; Heat Pumps and Infrastructure: $100,000 to $200,000; Total Investment: $1,000,000 to $1,550,000. They compared these costs to what would be paid just to be connected to the grid over a period of, say, 10 years.

Ultimately, the company's journey to energy autonomy did not begin with the installation of solar panels or the purchase of a battery system—it began with curiosity and a growing awareness of its impact on the environment prompted by energy costs both directly and indirectly related to grid reliability and the volatile cost of electricity. The journey began when the company received a shockingly high electricity bill during the winter months. Coupled with an increasing number of power outages and news of climate change, management realized they needed a more sustainable, reliable energy solution. The first step was education. The management team spent evenings researching renewable energy options. They soon discovered that microgrids offered both environmental and financial benefits. What appealed to them most was the promise of energy resiliency, especially during the frequent power outages their manufacturing facility experienced due to storms and an aging infrastructure.

The idea of energy autonomy and microgrid implementation is useful not just in homes, but also in businesses and industry. Businesses can use renewable energy sources and modern energy management systems to make the most of their energy, cut costs, and be more environmentally friendly. This section connects with the discussions in Chapter 2 on the technical foundations of microgrids and Chapter 3 on the economic considerations and business models. It shows how these principles can be applied in commercial contexts to achieve energy resilience and sustainability.

Foreseeing the Future of Energy Innovation

John Greenberg, a dedicated researcher at an institute dedicated to researching new energy technologies eagerly attends a scientific conference focused on the latest advancements in energy technology. As he immerses himself in the wealth of information and insights shared by experts from around the world, John is filled with excitement and anticipation for the future of energy innovation.

Throughout the conference, John learns about groundbreaking research and emerging technologies that promise to revolutionize the energy landscape. From advancements in renewable energy generation to breakthroughs in energy storage and microgrid technologies, the possibilities seem limitless.

Armed with newfound knowledge and inspiration, John returns to his institute with a renewed sense of purpose and determination. As he prepares to compile his findings into a memo for the company's leadership, he realizes that the task ahead will be easier than anticipated. The wealth of information and insights gleaned from the conference will not only inform the company's research and development efforts but also pave the way for new innovations and breakthroughs in the mid-term future.

John's excitement is palpable as he envisions a future where renewable energy, microgrids, and storage solutions are not just concepts but tangible realities, accessible to all. With continued investment in research and development, industries are poised to play a pivotal role in shaping this future, driving progress and innovation in the quest for a more sustainable and prosperous world.

John's Memo

In his memo following the scientific conference, John Greenberg highlights ten key innovations poised to transform the energy landscape.

1. **Next-generation Solar Panels:** Advancements in solar panel technology, such as perovskite solar cells and bifacial panels, promise higher efficiency and lower costs, making solar energy more accessible and widespread. For example, perovskite solar cells have shown potential to achieve higher efficiency rates compared to traditional silicon-based cells [5]. Additionally, bifacial panels can capture sunlight from both sides, increasing overall energy output [6].
2. **Advanced Energy Storage Solutions:** Breakthroughs in battery technology, including solid-state batteries and flow batteries, offer higher energy density, longer lifespan, and faster charging capabilities, enhancing the reliability and flexibility of energy storage systems. Solid-state batteries, for instance, are known for their safety and high energy density [7]. Flow batteries, which store energy in liquid electrolytes, provide scalable and long-duration energy storage solutions [8].
3. **Smart Grid Integration:** Innovations in grid management software and hardware enable the seamless integration of renewable energy sources, demand-side management, and grid-scale energy storage, optimizing grid efficiency and stability. Smart grids incorporate advanced metering infrastructure and real-time data analytics to enhance grid reliability and efficiency [9].
4. **Distributed Energy Resources (DERs):** The proliferation of DERs, such as rooftop solar, small-scale wind turbines, and residential energy storage systems, decentralizes energy production and consumption, empowering consumers and enhancing grid resilience. Examples include community solar projects and home battery storage systems [10].
5. **Internet of Things (IoT) in Energy:** The integration of IoT devices and sensors in energy infrastructure enables real-time monitoring, optimization, and predictive maintenance, improving energy efficiency and reducing operational costs. Smart meters and connected appliances are common IoT applications in the energy sector [11].
6. **Blockchain for Energy Transactions:** Blockchain technology facilitates transparent and secure peer-to-peer energy transactions, enabling decentralized energy markets and fostering greater autonomy and efficiency in energy trading. Peer-to-peer energy trading platforms allow consumers to buy and sell excess energy directly [12].
7. **Electrification of Transportation:** The electrification of transportation and charging infrastructure reduces reliance on fossil fuels and creates new opportunities for grid integration and demand-side management. Electric vehicles and charging networks are examples of this trend [13].
8. **Hydrogen Energy Economy:** Advancements in hydrogen production, storage, and fuel cell technology pave the way for a hydrogen-based energy

economy, offering a clean and versatile alternative to fossil fuels. Hydrogen fuel cells are used in various applications, from transportation to industrial processes [14].

9. **Advanced Materials for Energy Efficiency:** Novel materials, such as thermoelectric materials and phase-change materials, enhance energy efficiency in buildings, appliances, and industrial processes, reducing energy consumption and environmental impact. Thermoelectric materials can convert waste heat into electricity, improving overall energy efficiency [15].
10. **Artificial Intelligence (AI) in Energy Optimization:** AI algorithms and machine learning techniques optimize energy production, distribution, and consumption, improving system efficiency, reliability, and cost-effectiveness. AI is used in smart grids to predict energy demand and optimize energy distribution [16].
11. **Clean Jet Fuel Technology:** An article in *The Wall Street Journal* highlights the growing interest in clean jet fuel technology, which is gaining traction among investors [17]. This technology aims to produce sustainable aviation fuel (SAF) from renewable sources, significantly reducing the carbon footprint of air travel. Companies are developing processes to convert waste materials and biomass into jet fuel, offering a promising solution for the aviation industry's environmental challenges.

These innovations represent the cutting edge of energy technology, promising to revolutionize the way we produce, distribute, and consume energy in the years to come.

Factors Influencing a US-based Company's Decision to Adopt a Microgrid

Rising Energy Costs: With energy bills steadily increasing, a company might see a microgrid as a strategic way to control costs by producing its own power. For example, a company is alarmed by electricity rates having increased by 20% over the past decade. By generating its own electricity and utilizing net metering, it could potentially eliminate its electricity bills and exposure to future increases. If its most recent annual electricity cost was $250,000, the company could potentially save this amount in year one in previously incurred utility bills once the microgrid is commissioned, with savings increasing over time in direct positive correlation to the increases in cost of electricity.

Grid Reliability and Resilience: Frequent outages, particularly during extreme weather events, pushed the company to seek a more reliable solution. Microgrids can operate independently from the main grid, ensuring continuous power during outages. For instance, during a recent storm, a facility might experience a prolonged blackout, but a microgrid would have kept operations running smoothly throughout the grid outage.

Environmental Concerns: A company committed to sustainability is motivated by a desire to reduce its carbon footprint. By switching to renewable energy, it can significantly reduce its greenhouse gas emissions. For example, using solar energy instead of fossil fuels can reduce a company's carbon footprint by up to 80%.

Energy Independence: The appeal of being energy independent—generating, storing, and managing one's own electricity—can be a major incentive for a company to consider installing a microgrid. Companies can view microgrid technology as a way to take control of their energy future. This independence is crucial during natural disasters or grid outages when the main power grid may be compromised; a microgrid would continue to power essential operations, ensuring business continuity and safety. Also, the company's budget wouldn't be subjected to increasingly volatile and rising electricity prices, removing that element of risk and complexity when making long-term business plans.

Government Incentives: The availability of tax credits and rebates play a significant role in making a microgrid financially viable. For instance, US federal tax credits can cover up to 30% of the installation costs for solar energy systems. If the initial cost of the microgrid was $1,800,000, one could receive a tax credit of $540,000, reducing our out-of-pocket expense to $1,260,000. The federal solar tax credit, also known as the Investment Tax Credit (ITC), allows businesses to claim 30% of their solar system costs as a tax credit. This credit can be claimed in the same year the system is installed and can be spread out over several years if the full credit cannot be used in one year. [18]

To start the procedure for claiming the federal solar tax credit, businesses should follow these steps:

1. **Consult with a Tax Professional:** Before proceeding, it is advisable to consult with a tax professional to understand the specific implications for your business and ensure you meet all eligibility requirements.
2. **Install the Solar Energy System:** Ensure that your solar energy system is installed by a certified professional. Keep all receipts and documentation related to the installation costs.
3. **Complete IRS Form 3468:** This form is used to claim the Investment Tax Credit. You will need to provide details about the solar energy system and the total installation costs. The form can be found on the IRS website [19].
4. **File your Tax Return:** Include Form 3468 with your annual tax return to claim the credit. If the credit exceeds your tax liability for the year, you can carry the unused portion forward to future years.
5. **Stay Informed:** Keep up-to-date with any changes to the tax credit or additional incentives that may become available. The US Department of Energy and EnergySage provide valuable resources and updates on solar incentives [20].

By leveraging these government incentives, one can significantly reduce the financial burden of transitioning to a microgrid, making it a more attractive and feasible option for your company. Keep in mind this credit may be eliminated by legislation currently working its way through the US Congress.

By considering these factors, a company can make an informed decision to adopt a microgrid, balancing financial savings, reliability, environmental impact, and energy independence. A company should only embark on its path to energy autonomy after extensive research, consultations with experts and careful consideration of both the costs and long-term benefits.

Conclusion: The Greenberg Family's Journey to Energy Autonomy

As we conclude Chapter 5, the Greenberg family's journey towards energy autonomy serves as an example of how individuals and communities can achieve energy independence. This chapter has provided insight into the decision-making process, implementation strategies, and the benefits of becoming part of an energy community.

The Greenberg family's drive for energy independence began with a thorough assessment of their energy needs and goals. They explored various renewable energy options, including solar panels and battery storage systems, to create a sustainable and resilient energy solution. Their journey was marked by careful planning, collaboration with experts, and a commitment to sustainability.

Throughout the chapter, we followed the Greenbergs as they navigated the challenges of transitioning to energy autonomy. They faced obstacles such as regulatory hurdles, financial considerations, and technical complexities. However, their perseverance and dedication to their vision enabled them to overcome these challenges and achieve their goals.

The Greenberg family's experience highlights the importance of community involvement and collaboration in the energy transformation. By joining an energy community, they were able to share resources, reduce costs, and enhance social cohesion. Their story demonstrates how collective efforts can lead to greater energy resilience and sustainability.

Commercial applications of energy autonomy were also discussed, showcasing how businesses can benefit from renewable energy solutions. The Greenbergs' journey illustrated the potential for cost savings, increased reliability, and reduced environmental impact in commercial settings. This case study provided valuable insights into the practical aspects of implementing energy solutions in both residential and commercial contexts.

As we move forward, it is essential to recognize that the journey towards energy autonomy is an ongoing process. Continuous monitoring, evaluation, and adaptation are necessary to ensure the long-term success and sustainability of energy systems. By staying informed about the latest technological advancements and regulatory changes, individuals and communities can optimize their energy solutions and maximize their benefits.

With a comprehensive understanding of how energy autonomy can be achieved at the individual and community levels, we are now ready to conclude our exploration of energy independence. The final chapter, titled "Conclusion: The Path Forward," will summarize the key insights and lessons learned throughout the book.

Conclusion: The Path Forward

The journey towards energy autonomy is not just a technical or economic challenge; it is something we must do in a world where climate change, a lack of energy security, and economic inequality threaten our future well-being and livelihoods. By using microgrids and being self-sufficient in energy, we can create a more resilient, sustainable and fair energy ecosystem for many. This means that today's needs are met without preventing future generations from meeting their own needs.

The decentralization of energy resources marks a departure from the traditional energy paradigms that have dominated the past century. Centralized systems, while efficient in their time, have become increasingly strained under the concurrent pressures of increasing electrification of personal energy use, industrial innovation (e.g., advent of AI) and climate change. The vulnerabilities inherent in these systems—ranging from aging infrastructure and inefficiencies to their susceptibility to natural disasters, physical attacks and cyber attacks—have exposed the limitations of relying solely on a single, interconnected grid. Moreover, the environmental costs associated with large-scale fossil fuel consumption have prompted a global reassessment of our energy practices.

Microgrids offer a viable and compelling alternative. By decentralizing energy production and integrating renewable energy sources, microgrids can provide a cleaner, more reliable, and more resilient energy supply. They allow communities to tailor energy solutions to their specific needs, reducing dependence on external sources and increasing control over their energy futures. This shift is not just about mitigating risks; it is about seizing new opportunities for innovation, economic development, and social empowerment.

Throughout this book, we have explored the various dimensions of microgrids—technical, economic, social, and environmental. We have seen how microgrids can be designed to meet the unique needs of different communities, from urban centers seeking greater resilience to rural areas in need of electrification. The hypothetical and real examples presented demonstrate the diversity of microgrid applications and the positive impact they can have on local economies, public health, and the environment.

The economic benefits of microgrids are manifold. While the initial costs can be substantial, the long-term savings in energy expenditures, combined with the potential for generating revenue through the sale of excess renewable power, make microgrids an attractive investment. Furthermore, the ability to leverage local resources and labor in the construction and operation of microgrids can stimulate economic activity and job creation within the community.

From a social perspective, microgrids represent a democratization of energy. They empower individuals and communities to take control of their energy supply, reducing dependence on distantly owned and managed energy providers. This can lead to greater energy equity, particularly in marginalized or underserved areas. By providing reliable and affordable energy, microgrids can improve quality of life, enhance educational and economic opportunities, and contribute to the overall well-being of the community.

Environmentally, the shift towards microgrids is a critical step in the global effort to combat climate change. By reducing reliance on fossil fuels and increasing the use of renewable energy sources, microgrids can significantly lower greenhouse gas emissions. Moreover, the localized nature of microgrid systems reduces the environmental impact associated with large-scale energy infrastructure, such as transmission lines and power plants.

Looking to the future, the potential for microgrids is vast. Advances in technology, particularly in energy storage and smart grid systems, will continue to enhance the efficiency and viability of microgrids. As more communities and businesses recognize the benefits of energy autonomy, the demand for microgrid solutions is likely to grow. Governments and policymakers also have a crucial role to play in supporting the development of microgrids through favorable regulations, incentives, and investment in research and development. This means giving people reasons to invest, making the rules clearer and easier to follow, and spending money on research and development. This will help to overcome the current financial and technical barriers to the wider use of microgrids. However, the transformation to a decentralized energy system is not without its challenges. Regulatory hurdles, financial constraints, and technical complexities must be addressed to ensure that the benefits of microgrids are realized on a broad scale. Collaboration between the public and private sectors, as well as the active participation of communities, will be essential in overcoming these obstacles. It is also very important to consider using renewable energy sources like solar, biomass and wind power in microgrids. These technologies are a sustainable alternative to fossil fuels, reducing greenhouse gas emissions and mitigating the impacts of climate change. Energy storage has also improved and become much more affordable, which means that microgrids will be able to economically supply energy reliably even when the sun isn't shining or the wind isn't blowing.

The emergence of new economic models and business strategies show how microgrids can be a good investment. By using local resources, creating new ways to make money and getting the community involved, microgrids can help the economy to grow and create jobs. This is particularly important for communities that don't have much money, where having access to affordable and reliable energy can help with social and economic development.

Looking to the future, it is clear microgrids will be an important component of the global energy transformation. The continued development of smart grid technologies, along with advances in renewable energy and storage solutions, will drive the evolution of decentralised energy systems.

These innovations represent the cutting edge of energy technology, promising to revolutionize the way we power our lives. It is important to reflect on the broader implications of the shift towards energy autonomy and the pivotal role microgrids will play in shaping the future of our energy systems. The concept of energy self-sufficiency, as explored in this book, is not merely about technological innovation or economic strategy—it is a profound reimagining of how energy is produced, distributed, and consumed in a world that increasingly values sustainability, resilience, and fair access to electricity.

In conclusion, the journey towards energy autonomy is one of both necessity and opportunity. It is a response to the pressing challenges of our time—climate change, energy security, and economic inequality—but it is also a chance to reimagine and reshape our energy systems for the better. Microgrids offer a path forward, one that is aligned with the principles of sustainability, resilience, and social justice.

Future Challenges: Energy Autonomy's Bigger Picture

This book opens the door on the fascinating world of microgrids and energy autonomy. It explores the motivations, technologies, economic and social implications, and opportunities offered by this energy transition. However, the road to a decentralised and sustainable energy future is littered with challenges that require ongoing reflection and concerted action.

The energy sector is constantly evolving, and new technologies for producing, storing and managing energy are emerging rapidly. Microgrids must be designed to be flexible and scalable in order to adapt to these technological advances. The integration of AI, blockchain and the IoT into microgrids is promising, but it also raises questions of compatibility, security, and confidentiality. Microgrids, as interconnected and digital systems, are vulnerable to cyberattacks. Robust security measures are essential to protect the operational integrity of microgrids and ensure data security. Intrusion protection, data encryption, and user training are crucial elements of an effective cyber security strategy.

The proliferation of microgrids requires common standards and protocols to ensure interoperability between the different systems. Communication between microgrids, energy exchange, and coordination with the main grid must be harmonized for efficient energy management.

Existing regulations are often not adapted to the complexity and decentralization of microgrids. Flexible and innovative regulatory frameworks are needed to encourage the uptake of microgrids, facilitate interconnection with the main grid, and ensure fair competition in the energy market. The success of microgrids depends on the support and active participation of communities. Awareness campaigns, educational programs, and training initiatives are essential to inform the public about the benefits of microgrids, dispel misconceptions, and encourage the adoption of sustainable energy practices.

Large-scale deployment of microgrids requires significant investment and sustainable business models. Innovative financing mechanisms, such as public-private partnerships, participatory financing, and green bonds need to be explored to support the development of microgrids. The transition to renewable energies, while beneficial for the environment, raises questions about the recycling and waste management of end-of-life equipment such as solar panels and batteries.

Effective recycling and reuse solutions must be put in place to minimize the environmental impact of microgrids.

Microgrids must be designed to withstand the effects of climate change, such as extreme weather events and temperature variations. Adaptation strategies, such as diversifying energy sources, strengthening infrastructure, and integrating intelligent risk management systems, are crucial to ensuring the resilience of microgrids to climate challenges. Continued investment in research and development is needed to improve microgrid technologies, reduce costs, and develop new energy solutions. Innovation in the fields of energy storage, renewable energies, AI, and cyber security is essential to drive forward the field of microgrids and realize its full potential.

In conclusion, the transition to a decentralized and sustainable energy future, while essential, presents many complex challenges. By tackling these challenges with determination and innovation, one can create an energy system that meets your needs while also doing a part, even if small, in making the planet more livable for future generations. To attain this goal, the path forward requires close collaboration between governments, businesses, researchers, and communities, each playing a crucial role in building a more sustainable, resilient, and equitable energy future.

References

1. https://www.iea.org/commentaries/empowering-people-the-role-of-local-energy-communities-in-clean-energy-transitions.
2. https://eu.usatoday.com/story/news/investigations/2024/07/31/summer-heat-wave-electricity-bills-impact/74221050007/.
3. https://www.acq.osd.mil/eie/Downloads/IE/Metrics%20and%20Standards%20for%20Energy%20Resilience%2020%20May%202021.pdf.
4. Business Continuity. https://www.bcpcouncil.gov.uk/business/business-support-and-advice/business-continuity.
5. https://www.cleanenergyreviews.info/blog/latest-solar-panel-cell-technology.
6. https://metropolismag.com/viewpoints/the-next-generation-of-solar-energy/.
7. https://www.solarpowerworldonline.com/2022/09/the-rise-of-next-generation-n-type-solar-pv-cells/.
8. https://www.technologyreview.com/2024/01/08/1085124/super-efficient-solar-cells-breakthrough-technologies/.
9. https://theconversation.com/new-solar-cells-break-efficiency-record-they-could-eventually-supercharge-how-we-get-energy-from-the-sun-239417.
10. https://www.hivepower.tech/blog/smart-grid-technology-changing-the-power-industry.
11. https://en.wikipedia.org/wiki/Smart_grid.
12. https://intelliarts.com/blog/smart-grid-iot/.
13. https://www.energy.gov/eere/analysis/assessing-value-electric-vehicle-managed-charging-review-methodologies-and-results.
14. https://www.mckinsey.com/industries/electric-power-and-natural-gas/our-insights/how-grid-operators-can-integrate-the-coming-wave-of-renewable-energy.

15. https://powerefficiency.com/advanced-energy-storage-technologies/.
16. https://www.startus-insights.com/innovators-guide/applications-of-energy-storage/.
17. https://www.wsj.com/business/the-clean-jet-fuel-technology-winning-over-wall-street-f20eac23?st=y1UL8R&reflink=article_whatsapp_share.
18. https://solaramerica.com/insights/2024-federal-solar-tax-credit.
19. https://solaramerica.com/insights/2024-federal-solar-tax-credit.
20. https://www.energy.gov/eere/solar/homeowners-guide-federal-tax-credit-solar-photovoltaics.

Glossary

Aggregator: An entity that groups together multiple distributed energy resources (DERs) to act as a single entity in energy markets. Aggregators can optimize the collective output and provide grid services.

Ambient Heat: Natural heat present in the environment (in air, water, soil) can be utilized for energy production.

Analysis Systems: These are systems that process the data transmitted by communication devices and optimize the management of the electrical network. They use advanced algorithms, such as artificial intelligence, machine learning, and optimization. They take into account the technical, economic, and environmental characteristics and objectives of the network.

Ancillary Services: Support services necessary to maintain grid stability and reliability, such as frequency regulation, voltage control, and spinning reserve. Microgrids can provide ancillary services to the main grid.

Autonomy: The ability of an energy system, particularly a microgrid, to operate independently of the central electrical grid. Autonomy ensures energy self-sufficiency and reduces reliance on external energy suppliers.

Battery Energy Storage System (BESS): A system that stores energy generated from renewable sources like solar or wind for later use. Batteries are key components in microgrids, providing power when the sun isn't shining or the wind is not blowing.

Biomass Energy: Energy produced from organic materials, such as plant and animal waste. Biomass can be used for heating, electricity generation, and as a fuel for transportation.

Black Start Capability: The ability of a power system, such as a microgrid, to restart independently after a total shutdown without relying on external power sources.

Blockchain for Energy Transactions: A decentralized ledger technology used for transparent and secure peer-to-peer energy trading. This enables prosumers to sell excess energy directly to neighbors or other users without an intermediary.

Capacity Factor: The ratio of actual energy produced by an energy system to the maximum possible energy it could produce over a given period. Higher capacity factors indicate more efficient energy production.

Carbon Footprint: The total amount of greenhouse gasses (GHGs) emitted directly or indirectly by human activities, including energy production from fossil fuels.

Carbon Neutrality: Achieving a balance between emitting carbon and absorbing carbon from the atmosphere. This can be accomplished by using renewable energy sources and carbon offset practices.

Circular Economy: An economic system aimed at minimizing waste and making the most of resources. For energy systems, this involves recycling components like solar panels and batteries at the end of their lifecycle.

Combined Heat and Power (CHP): A system that simultaneously generates electricity and useful heat from the same energy source. CHP systems improve overall energy efficiency by utilizing waste heat.

Communication Devices: In the context of electrical networks, these are devices that transmit the data collected by the sensors and the orders sent to the actuators. They use different means of communication, such as radio waves, optical fibers, and wireless networks.

Community Microgrid: A localized energy grid shared by multiple households or businesses. These microgrids provide energy autonomy, security, and resilience to participating members.

Cybersecurity: Measures taken to protect energy systems, particularly decentralized microgrids, from cyberattacks or unauthorized access. As energy systems become more digitized, cybersecurity is critical for operational integrity.

Decarbonization: The process of reducing or eliminating carbon dioxide emissions from energy production by transitioning to low-carbon energy sources such as wind, solar, and hydropower.

Decentralization: The shift from large, centralized power plants to local, distributed energy generation systems like microgrids. Decentralization reduces transmission losses and enhances grid resilience.

Demand Response (DR): A system of adjusting energy demand based on supply conditions. Demand response can lower consumption during peak hours and help balance the grid.

Demand-Side Management (DSM): Strategies and technologies used to control and reduce energy consumption on the demand side. DSM includes energy efficiency measures, demand response, and load shifting.

Digitalization: The use of digital technologies like sensors, smart meters, and AI to optimize energy production, distribution, and consumption in real-time. Digitalization is crucial for smart grids and microgrids.

Distributed Energy Resources (DERs): Small-scale energy generation or storage technologies located close to where energy is used. DERs include solar panels, wind turbines, batteries, and electric vehicles.

Energy Autonomy: The ability to produce, store, and manage energy independently, without relying on the central grid. Energy autonomy often involves using renewable energy sources and battery storage.

Energy Community: A group of prosumers who collectively manage energy resources like solar panels and storage, often through a shared microgrid. Energy communities focus on self-sufficiency, sustainability, and resilience.

Energy Equity: Ensuring fair access to clean and affordable energy for all, especially marginalized and lower-income communities. Energy equity seeks to address disparities in energy access.

Energy Management System (EMS): A system that monitors, controls, and optimizes the performance of energy resources like solar panels, batteries, and appliances to ensure efficient energy use.

Energy Poverty: The lack of access to reliable and affordable energy, often affecting low-income or remote communities. Energy poverty can be mitigated through decentralized energy systems like microgrids.

Energy-as-a-Service (EaaS): A business model where customers pay for energy services rather than owning and maintaining energy infrastructure. EaaS providers manage energy generation, storage, and optimization.

Feed-in Tariff (FiT): A policy mechanism that guarantees a fixed payment for renewable energy producers for the electricity they generate and feed into the grid. FiTs encourage investment in renewable energy.

Geothermal Energy: Obtained by harnessing heat from the Earth's interior for electricity production or direct heating.

Green Hydrogen: Hydrogen produced using renewable energy sources, such as electrolysis powered by solar or wind. Green hydrogen is a clean fuel with potential applications in transportation, industry, and energy storage.

Grid Parity: The point at which renewable energy becomes as cost-effective as traditional fossil fuels. Achieving grid parity is a major milestone in the transition to renewable energy.

Hybrid System: An energy system that combines multiple energy sources, such as solar and wind, with energy storage to provide a reliable and stable power supply.

Hydropower: Derived from the force of water, often harnessed by dams, hydraulic turbines, or other devices.

Interconnection Agreement: A contract between a microgrid or distributed energy resource and the main grid operator, outlining the terms and conditions for connecting to the grid.

Inverter: A device that converts direct current (DC) electricity generated by solar panels into alternating current (AC) electricity, which is used in most homes and businesses.

Island Mode: The ability of a microgrid to disconnect from the main grid and operate independently during outages or disruptions. This mode enhances grid resilience and reliability.

Levelized Cost of Energy (LCOE): A measure of the average cost of producing electricity from a specific energy source over its lifetime, accounting for capital, operating, and maintenance costs.

Life-Cycle Analysis (LCA): A process that assesses the environmental impacts of a product or system, such as solar panels or batteries, from production to disposal. LCAs help evaluate the sustainability of energy technologies.

Load Balancing: The process of distributing energy demand across different times to avoid peak loads and reduce stress on the grid. Load balancing is essential for the stability of microgrids and smart grids.

Microgrid Controller: A device or software that manages the operation of a microgrid, including energy generation, storage, and distribution. The controller ensures optimal performance and stability.

Microgrid: A small, self-contained energy system that can generate, store, and manage energy locally. Microgrids can operate independently of the main grid, providing energy resilience and sustainability.

Net Metering: A billing mechanism that credits solar energy system owners for the electricity they add to the grid. Net metering encourages the adoption of renewable energy by providing economic benefits to prosumers.

Net Zero Energy Building (NZEB): A building that produces as much energy as it consumes on an annual basis, typically through the use of renewable energy sources and energy efficiency measures.

Off-Grid System: An energy system that operates independently of the main electrical grid, typically relying on renewable energy sources and energy storage systems like batteries.

Peer-to-Peer (P2P) Energy Trading: A decentralized energy market where prosumers can buy and sell energy directly with each other using digital platforms and blockchain technology.

Perovskite Solar Cells: A new generation of solar cells known for their high efficiency and potential to reduce the cost of solar power. Perovskite technology is an emerging field in renewable energy.

Power Purchase Agreement (PPA): A contract between an energy producer and a buyer, typically used in large-scale renewable energy projects. PPAs can also be applied in microgrid partnerships between businesses and communities.

Prosumer: A person or entity that both produces and consumes energy. Prosumers play a central role in decentralized energy systems by generating their own energy and potentially selling excess back to the grid.

Public-Private Partnership (PPP): A collaboration between government entities and private companies to develop energy infrastructure, such as microgrids. PPPs often help finance large-scale renewable energy projects.

Renewable Energy: Energy derived from natural sources that replenish themselves, such as solar, wind, hydro, and geothermal energy. Renewable energy is a cornerstone of sustainable energy systems.

Renewable Portfolio Standard (RPS): A regulatory policy that requires a certain percentage of electricity to come from renewable energy sources. RPS policies drive the adoption of renewable energy.

Resilience: The ability of an energy system to withstand and recover from disruptions, such as extreme weather events or cyberattacks. Microgrids enhance resilience by localizing energy generation and storage.

Sensors: Devices that measure the parameters of the electrical network, such as electricity flows, consumption levels, current quality, etc. They allow the collection of real-time data on the state of the network and the detection of anomalies or malfunctions.

Smart Grid: An advanced electrical grid that uses digital technology to monitor and manage the flow of electricity. Smart grids allow for more efficient energy distribution, demand response, and integration of renewable energy.

Smart Inverter: An advanced inverter that can communicate with the grid and adjust its output to support grid stability. Smart inverters are essential for integrating high levels of renewable energy.

Smart Meter: A digital device that records real-time electricity consumption, allowing for better energy management. Smart meters are essential in optimizing energy use in microgrids and smart grids. They promote interaction between the network and consumers and encourage ecologically responsible behaviors.

Solar Energy: Generated through the conversion of sunlight into electricity or heat, typically using photovoltaic or thermal solar panels.

Solar Home System (SHS): A standalone solar power system designed for homes, typically in remote areas. SHSs provide energy independence by using solar panels and batteries to meet household electricity needs.

Sustainability: The practice of using resources in a way that meets current needs without compromising future generations' ability to meet theirs. Sustainability is a core principle in renewable energy systems.

Time-of-Use (TOU) Pricing: A pricing scheme where electricity rates vary based on the time of day, encouraging consumers to shift their energy use to off-peak periods.

Vehicle-to-Grid (V2G): A technology that allows electric vehicles to discharge stored energy back into the grid, providing additional energy storage and grid support.

Virtual Net Metering (VNM): A policy that allows multiple energy consumers to share the benefits of a single renewable energy system, such as a community solar project.

Virtual Power Plant (VPP): A network of decentralized energy resources, such as solar panels and batteries, that are integrated and managed as a single power plant through digital platforms.

Voltage Regulation: The process of maintaining a consistent voltage level on the electrical grid. Voltage regulation is crucial for grid stability and the safe operation of electrical devices.

Watt (W): The unit of power in the International System of Units (SI), representing the rate of energy transfer equivalent to one joule per second. It is commonly used to measure electrical power.

Watt-hour (Wh): A unit of energy representing one watt of power expended for one hour. It is commonly used to measure electricity consumption.

Wave and Tidal Energy: Extracted from the kinetic energy of waves or tides to generate electricity.

Wind Energy: Produced by capturing the kinetic energy of the wind through wind turbines to generate electricity.

Zero Emission Vehicle (ZEV): A vehicle that produces no tailpipe emissions, typically powered by electricity or hydrogen fuel cells. ZEVs contribute to reducing air pollution and greenhouse gas emissions.

Zero Net Energy (ZNE): A building or system that produces as much energy as it consumes over a year, typically using renewable energy sources like solar panels.

Index

A

Advanced Metering Infrastructure (AMI) 15, 23, 181
Ancillary Services 50
Artificial Intelligence (AI) 1, 8, 118, 182
Autonomy 1, 2, 19-22, 26, 27, 30, 35, 43-46, 57, 78, 79, 85, 92, 99, 100, 113, 118, 119, 125, 126, 161, 165, 168, 170, 174, 176, 180, 184, 186

B

Battery Energy Storage System (BESS) 103, 104, 124
Battery Recycling 65
Behind-the-Meter (BTM) 93
Bioenergy 31
Biomass Energy 32, 105
Blockchain for Energy Transactions 181

C

Carbon Footprint 19, 46, 48, 57, 58, 61, 76, 106, 113, 118, 133, 134, 136, 138, 160, 161, 165, 167, 169, 173, 174, 176, 179, 182
Carbon Neutrality 30, 140, 141, 145
Circular Economy 60, 113, 118, 133, 135, 136, 161
Climate Mitigation 21
Combined Heat and Power (CHP) 53, 102
Community Microgrid 95, 96, 104, 119
Concentrated Solar Power (CSP) 69
Critical Infrastructure 127
Cybersecurity 25, 85, 102, 131, 132

D

Decarbonization 19, 31, 61, 148, 149
Decentralization 148, 149, 154, 185, 187
Demand Response (DR) 23, 49, 78, 102, 127, 128, 129, 143, 147, 151, 154, 156, 165
Demand-Side Management (DSM) 48, 175, 181
Digital Twin 127 129, 161
Digitalization 11, 126, 148, 149
Distributed Energy Resources (DER) 23, 46, 50, 181
Dynamic Pricing 51, 52, 157

E

Electric Heat Pump 127, 154
Electric Vehicle (EV) 14, 18, 30, 51, 64, 67, 106, 179, 181
Electric Vehicle Charging Station 106
Electricity Market 8, 22, 52, 139
Electrolyzer 63
Energy Autonomy 1, 2, 20, 26, 30, 45, 46, 48, 57, 58, 78, 79, 85, 87, 92, 113, 114, 118, 119, 124, 125, 133, 139, 161, 162, 165, 167, 168, 170, 175, 180, 184, 186, 187
Energy Blockchain 181
Energy Communities 9, 19, 96, 111, 121, 126, 133, 146, 148, 149, 155, 178
Energy Efficiency 1, 23, 37, 48-50, 55, 56, 58, 74, 76, 93, 94, 97, 101, 114, 124, 140, 141, 143, 161, 171, 175, 181, 182
Energy Equity 1, 185
Energy Independence 2, 9, 21, 26, 30, 46, 48, 58, 59, 72, 75, 79, 95, 105, 111, 113, 119, 124, 140, 141, 145, 149, 150, 155, 156, 159, 162, 165-169, 172, 173, 175, 178, 179
Energy Intensity 56
Energy Justice 118
Energy Literacy 112
Energy Management System (EMS) 30, 40, 41, 121, 124, 158, 168, 180
Energy Mix 32, 60-62, 64
Energy Poverty 25, 42, 109, 110
Energy Security 11, 20, 22, 98, 104, 106, 128, 136, 142, 158, 165, 170, 176, 185, 187
Energy Sovereignty 104
Energy Transition 19, 60, 93, 94, 97, 118, 121, 126, 131, 187
Energy-as-a-Service (EaaS) 102

F
Feed-in Tariff (FiT) 89
Fuel Cell 9, 53, 71, 79, 88, 181, 182

G
Geothermal Energy 62
Green Hydrogen 60, 63, 71
Greenhouse Gas (GHG) 1, 16, 32, 40, 46, 48, 50, 52, 57, 60, 61, 63, 74, 76, 103, 136, 140, 142, 143, 158, 169, 183, 186
Grid Flexibility 19, 142
Grid Modernization 11, 141
Grid Stability 23, 50, 67, 110, 128, 142, 154, 156, 174

H
Home Energy Management System (HEMS) 40, 41, 158
Hybrid System 38, 40, 61, 67, 101, 107
Hydropower 31, 33, 34, 53, 62, 97, 149

I
Infrastructure Resilience 91
Inverter 17, 31, 38, 45, 75, 88, 110, 119, 134, 135, 172
IoT for Energy 181
Island Mode 43, 173

L
Load Balancing 6, 30, 128
Load Forecasting 128, 165
Low Carbon Economy 94, 142

M
Microgrid 1, 2, 8, 9, 12, 14-16, 21, 25, 35, 41, 52, 87, 95, 104, 110, 118, 128, 134, 151, 165, 170, 180, 186
Microgrid Controller 53, 106

N
Net Metering 21, 85, 91, 93-95, 172, 182
Net Zero 19, 90, 97, 140

O
Off-Grid System 35, 37, 79, 146
Offshore Wind 62
Onshore Wind 105

P
Peak Demand 6, 23, 52, 67, 102, 126, 139, 146, 166, 174, 179
Peer-to-Peer (P2P) Energy Trading 181
Perovskite Solar Cells 181
Photovoltaics (PV) 34, 35, 99
Power Purchase Agreement (PPA) 79
Predictive Maintenance 131, 181
Prosumer 1, 19, 23, 24, 109, 113, 118, 120, 121, 124, 125, 128, 131, 133, 139, 143, 145, 146, 148, 149, 166, 170
Public-Private Partnership (PPP) 85

R
Renewable Energy 8, 11, 12, 15, 16, 33, 36, 39, 44, 51, 58, 59, 62, 71, 74, 95, 101, 110, 128, 140, 154, 169, 180, 186
Resilience 1, 2, 103, 113, 118, 129, 132, 147, 174, 182
Retrofit 56, 97
Rooftop Solar 23, 46, 90, 102, 149, 170, 181

S
Self-Consumption 43-45, 48, 179
Smart Appliances 10, 157
Smart Grid 11, 13, 14, 30, 40, 49, 51, 67, 78, 106, 109, 111, 113, 119, 122, 126, 141, 186
Smart Home 11, 40, 58, 126
Smart Meter 14, 16, 17, 23, 51, 109, 126, 130, 149, 169, 171, 175, 181
Solar Farm 13, 75, 105, 135, 176
Solar Home System (SHS) 37, 38
Solar Thermal 61
Storage Capacity 23, 66, 67, 70, 85
Sustainability 1, 2, 8, 14, 23, 26, 32, 38, 48, 50, 78, 79, 105, 110, 113, 118, 121, 125, 131, 133, 135, 147, 150, 156, 161, 169, 173, 175, 184, 187
Sustainable Development Goals (SDGs) 114, 118, 142

T
Thermal Energy Storage 69
Time-of-Use (TOU) Pricing 51

U
Urban Microgrid 106

V
Vehicle-to-Grid (V2G) 118
Virtual Power Plant (VPP) 21, 27, 92, 155, 156

Biography

Emmanuel Fragnière is a Professor of Service Innovation and Risk Management at the Institute of Tourism, School of Management Valais-Wallis (HES-SO), Switzerland. He holds a Licence in Econometrics and a PhD in Management Science from the University of Geneva, with further studies at EPFL and Stanford University. His career bridges academia and industry, beginning at HEC Lausanne and continuing through roles in the financial and commodity sectors at Banque Cantonale Vaudoise and Cargill. For the past two decades, he has served as a professor within HES-SO and taught for many years at the University of Bath. His work contributes to advancing sustainable and resilient service systems, particularly in tourism and energy. He has published in leading journals including Management Science, European Journal of Operational Research (EJOR), Annals of Operations Research, Annals of Tourism Research Empirical Insights, and Digital Transformation and Society. He is also involved in EU projects on the social acceptance of micro smart grids (e.g. o-cei.eu) and serves as a RESER Council Member (European Association for REsearch on SERvices).

George Sullivan is a published and well-recognized subject matter expert in federal US Government contracting and procurement, having worked for both the Government for 10 years as a contracting specialist and officer in defense and non-defense agencies and over seven years for private industry as a contracts and compliance manager and director for public, private, and non-profit corporations in both the defense and energy sectors. He is an NCMA certified federal contract manager and licensed to practice law in California and Florida. He holds a JD from Florida International University (cum laude - 2008), and MBA from Thunderbird Graduate School of Global Management (2003), and BA in Russian Studies from UCLA (1996). He currently resides with his wife and two daughters in Northern Virginia.

Loïc Viret is a Swiss entrepreneur with solid international experience in industry and renewable energy. As a specialist in energy storage, he is currently developing large-scale projects, notably in collaboration with several electricity distributors. His technical background is complemented by an MBA (with a specialization in risk analysis) and a degree in educational training. He regularly lectures at EPFL and HES and leads masterclasses on energy self-sufficiency.

In 2023, he received the Industry 4.0 Award "The Shapers", a distinction awarded to individuals and companies in French-speaking Switzerland for their contribution to the advancement of innovation in Switzerland.

For Product Safety Concerns and Information please contact our EU representative GPSR@taylorandfrancis.com Taylor & Francis Verlag GmbH, Kaufingerstraße 24, 80331 München, Germany